Elements of
system-dynamics simulation

A textbook with exercises

Elements of system-dynamics simulation.

A textbook with exercises

Th. J. Ferrari

A HALSTED PRESS BOOK

JOHN WILEY & SONS
New York-Toronto

Published in the U.S.A., Canada and Latin America by Halsted Press, a Division of John Wiley & Sons, Inc., New York

Ferrari, Th J
Elements of system-dynamics simulation.

''A Halsted Press book.''
Bibliography: p.
Includes index.
1. System analysis. 2. Feedback control systems.
I. Title.
QA402.F47 003 78-10505
ISBN 0-470-26548-5

Printed in the Netherlands

Contents

Preface

This book is the revised content of various courses that the author has given in recent years to groups of persons from different scientific disciplines. Among them were students of economics and social science at the University of Groningen, students of environmental health at the University of Amsterdam, participants in courses for scientific co-workers of the Ministry of Agriculture and Fisheries at Wageningen and participants in courses of graduate scientists on Dynamic Simulation in Ecology at Wageningen.

All participants had in common the need to learn practical applications of the system-dynamics approach. The course also aimed at giving the participants an understanding of the origin of actual phenomena and the skill in converting the problems into system-dynamics relational diagrams and into simple algebraic equations. In planning the course, I assumed that most participants had little knowledge of mathematics and it served little purpose to bring it up to scratch. The same applies to the book. To understand it, one need know nothing of differential and integral calculus. The essentials are introduced gradually.

The course emphasized the practical application of system dynamics as is illustrated by the numerous exercises. In my experience, students can learn the concepts and principles of system dynamics fairly easily but have difficulty in translating and converting the problem into the language of system dynamics. Perhaps traditional approaches and accepted models of the various sciences have not paid sufficient attention to changes with time, i.e. to behaviour. In the course, theory and application were linked by setting problems for homework. The solutions were then discussed in the next lecture. Within 20–25 lectures, participants became sufficiently acquainted with system dynamics to solve their own problems.

Forrester and his school have aroused much interest in the application of system dynamics. There is, however, a need for a book dealing with its principles, the solution of models based on them, and the derivation of equations that describe these models more extensively. Such a book should relate theory to application. Therefore, I have devoted much attention in this book to problems and their solution.

The book starts with the discussion of systems and models, and then deals with the concepts of the system-dynamics approach. The reader must first study and become familiar with this part. He can test how well he has understood the text by answering various questions. Then come the exercises, which form the key to the whole work. Answers to both questions and exercises are included, so that the reader can check his own work. It is essential that the reader works out the problems himself! The exercises become more difficult as more phenomena are introduced but they are fairly simple, so that the reader should still see the wood for the trees. The text progresses from simple to more complex structures, so that an understanding is gained of the origin and essentials of some characteristic patterns of behaviour.

The book also explains how to set down the numerical solution of the differential equations with just pencil and paper. It does not deal with the use of a computer in such calculations, even though the computer is indispensable for the study of models, even simple ones. That task is outside the scope of this book: it is not a computer course. Moreover the value of training in simulation techniques is debatable, because the introduction of 'simulation languages' considerably simplifies the programming of such calculations.

I have dealt with patterns of behaviour that occur in different disciplines. Examples are taken from physics, chemistry, process control, biology, ecology, hydrology, food science and economics, although most examples are from natural science. The exercises are so treated that readers from other disciplines can solve them with system dynamics. Thus the reader becomes aware of how problems from various disciplines can be described and explained analogously. It is also good experience to draw abstractions from the real world. All this helps in an interdisciplinary approach to important problems for human welfare.

This book is intended for students, researchers and others interested in how feedback systems work. The reader must never lose sight of the fact that the usefulness of system dynamics depends on the user's knowledge of his own discipline. System dynamics is no answer to lack of expertise! However system-dynamics models can indicate where knowledge is lacking.

Finally, let me thank all those who by their interest and criticism have contributed to the content of the course and this book. Participants have played a major role in working out the course. I hope that their influence can be seen. The exercises are mainly system-dynamics elaborations of subjects from literature. Some exercises on business management are the work of J. V. Ferrari.

Groningen, September 1977

1 Introduction; systems and models

Nowadays, more than ever before, well-meant human action at a certain place and at a certain moment may have large and unexpected results, unfavourable for human welfare at another place and at another time. There is much knowledge about separate parts of a *system*, but it is difficult or even impossible to connect these pieces of information. Reality shows a behaviour which is characteristic for the system, of which the behaviour is more than the behaviour of the sum of the separate parts.

What is meant by the behaviour of a system is illustrated in Fig. 1, which shows the calculated behaviour of a biological system. It relates the interaction between two insect species: the parasites living off the host insects and the hosts killed by the parasites (Exercise 29). The numbers of both species show systematic oscillations comparable with those in nature. However, the most remarkable aspect of these oscillations is the way in which they originated. The initial condition before the calculations started was a *steady state* or *state of dynamic equilibrium* in which the numbers of both species were constant. In such a state, we introduced small randomized changes, as external noise, in the natural birth and death rates so that after some time large systematic oscillations were produced. The influences introduced were at random; therefore, it would be pointless to relate the characteristic attributes of these oscillations to just any environmental factor. Ele-

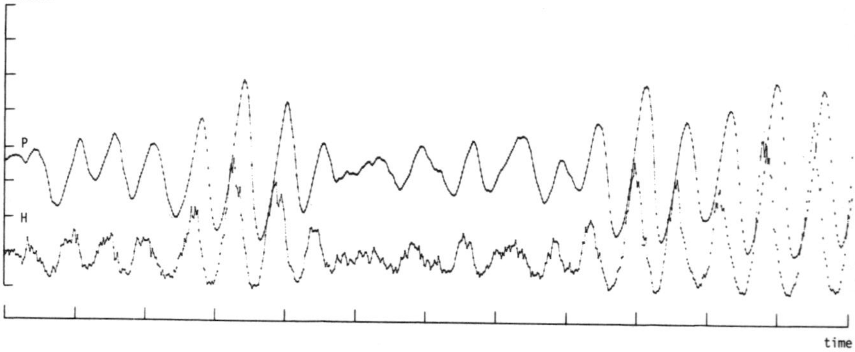

Fig. 1. Calculated behaviour of a system representing the interaction of two insect species. The system was affected only by external random noise.

ments in this system and its structure are responsible for this oscillatory behaviour. Such phenomena, which indeed are known in other disciplines such as economics and social science, point to the importance of the systems approach.

The use of systems is a rather recent development in scientific methodology. The first step was taken in biology. Further development in other disciplines has been more or less independent and has resulted in a lack of uniformity in concepts and in methods of systems thinking, also known as system analysis, system dynamics, systems theory and systems engineering. However uniformity is gradually emerging, so that the systems approach is becoming the connecting link between the different disciplines.

The concept *system* can be defined as a collection of associated *elements* sharing a common purpose or function. In this context, element is taken as given and is not investigated further. The element itself can be considered as a system too; then we increase the *level of generalization.* A sociological unit is a system which may be composed of the elements man, animal or plant; at a higher level of generalization man, animal or plant can be taken as a system. If there are relationships between the elements, their behaviour is no longer independent: a change in one element causes a change in one or more other elements in the system. Really the definition of a system is too general. We can only use it to distinguish systems from each other by a number of elements and relationships, by the *structure*, and consequent differences in *behaviour*, i.e. the changes in the state of the system during a certain period.

Actual systems can be classified according to different aspects. As the number and content of the relationships is important, a classification can be based on an increasing complexity of structure. For instance, Boulding (1956) introduced the following scheme:

Level 1 is characterized by a static structure, in which the time element is absent: examples are some physical laws, a map.

Level 2 has a dynamic structure with the time element: simple machines, a clock.

Level 3 includes control mechanisms, characterized by information transfer and processing: a thermostat.

Level 4 refers to self-maintaining systems. It is the level of the cell; it is the transitional state between inanimate material and life: virus, bacterium.

Level 5 is the start of functional separation: the plant.

Level 6 is characterized by movement and consciousness: the animal nature.

Level 7 is the level of the human being who is able to think.

Level 8 is related to social organizations: family, group, society.
Level 9 is less important for our purpose, it refers to the transcendental master-system.

An increase in complexity usually means an increase in *relaxation time* or *response time.* The former is the time needed to settle a system, unbalanced by an interference. For instance, if we compare the values for the restoration time of chemical processes, those for reactions of biological cells and those for behaviour of plant and plant communities, we see that they range from seconds up to years. In the discussion on the time coefficient, we will revert to this point. Reversely, with increasing complexity, it also takes more time, i.e. response time, to realize certain changes. The ideals of a single person are changed more quickly and enduringly than those of the group to which the person belongs.

According to Boulding's classification, each level includes the characteristics of the systems of lower levels. Thus we may incorporate the characteristics of a system of a high level in a system of a lower level, if necessary. In scientific research and management decisions, this procedure is often used in reverse. However, one then runs the risk of omitting essential characteristics of these higher-level systems and of not understanding how these systems work. Most models have this drawback.

Man has always used a *model* to vizualize the real world. Especially in the empirical sciences, models have caught on. They can be used to predict the effects of certain decisions and interferences. In this book we shall restrict ourselves to the following definition: a model is a representation of a system by a form different from that of the system represented. We shall only meet models in an analogous form (relational diagrams) and in a mathematical form (differential, finite-difference and other equations).

A model is applied especially in communication, instruction, planning and when forming hypotheses. Reality, however, is too complex to be represented completely so that a model has merits and significance only if all relevant elements and relationships have been included in the model. The question remains whether it is possible to represent satisfactorily the different systems according to Boulding's scheme.

The representation must fulfil the needs of *reality* and of *generality.* Reality is the extent to which the elements and the relationships, taken-up in the model, correspond to those in the real world. Generality relates to the number of situations and systems to which the model can be applied and is bound up with the number of elements and relationships in the model: the degree of *resolution* or *reticulation*, or reversely, the degree of *aggregation*.

An analysis of the historical development of model making shows why systems thinking did not develop until the last decade. The problems of system dynamics only appear with the representation of systems belonging to Level 3 and higher of Boulding's classification. In the beginning of the research, insight was still insufficient to feel the need to mimic such systems. A combination of circumstances and developments has undoubtedly contributed to the enormous increase in systems thinking during the last years: the increase in knowledge on control and servo mechanisms, the importance of the feedback loop for the behaviour of these technical systems, the working of decision functions and the significance of information for it. At the same time we have gained insight into the applicability of all this knowledge to the explanation of the behaviour of natural systems. Finally, the introduction of the computer has made it possible to solve the differential equations needed to describe feedback mechanisms and to make the models of such dynamic systems operable.

As a result, the technique of *simulation* which has been in use for many years, has gained new impetus. For a long time, emphasis in research has been on experimentation with the real world, with and without interference (Ferrari, 1965). During the last decades, the simulation technique has been used more and more in research (Shannon, 1975). Simulation means experimentation with models by changes in elements or in relationships in order to understand better the behaviour of the system or to compare the meanings and values of the different strategies. The simulation technique should play a great part in research and in management, especially with the development of systems theory. Simulation is often the only way to assess the extent of knowledge, for example when the mathematical or analytical solution of the model is too complicated. Often it is impossible to study real phenomena in isolation according to the ceteris-paribus principle, in which all factors, except one or two, are kept constant. In other situations, such isolation may give unrealistic results. Moreover, simulation is an outstanding method when experimenting with the real world is unethical as in social science, ecology and space flights or would take too long as in economics and ecology, or involves risks as in ecology, medical science and social science.

Which models have been used in research and in management up to now? To answer this question, we do not need to differentiate between the various sciences, as the sequence of development has always been the same. We are concerned here with models intended to represent the systems of the several levels of Boulding's scheme.

Although at the start of the development of the various sciences, models were applied in which the time variable was used explicitly, mostly static models (i.e. without time as variable) with one equation are still used in all sciences. The well-known regression equation is based on this model; it has the form $y = f(x_1, x_2 \ldots x_n)$ in which y is the dependent variable, x the independent or explanatory variables, and f represents some function of the independent variables. The use of this type of equation means that a structure of Level 1 is taken to represent systems of higher levels of complexity. Without underestimating the value of such a description, we may conclude that reality was poorly mimicked in many cases. However, many phenomena are still represented in this way! An improvement has been the introduction of models described not by one equation but by more equations. Such models assume that the dependent variables have an influence on other variables. The mathematical form of an arbitrary model could be: $y_1 = f_1(x_1, x_2, y_2, y_3)$, $y_2 = f_2(x_1, x_2)$ and $y_3 = f_3(x_2, x_3)$; in the first equation, the dependent variables y_2 and y_3 are now assumed to be explanatory variables. The model of the factor analysis belongs to this type too; the number of relationships, however, is assumed to be unlimited. All these models are static; one is interested in number and relative influences, the behaviour of the system as such is not considered.

From the beginning scientists have tried to introduce the dynamic character of a phenomenon into models. Often the rate of change in a variable can be expressed in a differential equation as a function of a number of factors. This kind of model is only useful if the change or behaviour can be inferred as a function of time by integration of the differential equation. This procedure has always been used in physics and chemistry and less frequently in other sciences. Naturally, scientists have tried to develop models with one or more differential equations, also for complex systems as in biology. However, this approach could hardly develop because it was impossible to solve or integrate analytically these more realistic and consequently more complex models. The computer and the development of systems thinking changed this situation.

The application of static models strongly stimulated the idea of causality in one direction; a change in the independent or causal variables only implies a change in the dependent variables. This change in a dependent variable does not influence the state of the independent one, neither now nor later. This way of scientific thinking has strongly determined all actions in science and in management. However, the development towards the systems-thinking approach has brought the causal loop more to the fore: each change in a state variable again influences the rate of change. It is clear that the scope of the system description is widened by the introduction of these 'feedback loops' into models of systems of Level 3 and higher in Boulding's scheme.

By using the ideas mentioned and the possibilities of the computer, Forrester (1961, 1968) put forward a theory about the content and structure of systems and how they work. To this, he coupled the numerical solution technique of differential equations. By using an adapted relational diagram technique, he introduced to non-specialists a way to apply this systems thinking. Forrester called this combination of theory and method *system dynamics*. It became well known through the publications of the Club of Rome (Meadows, 1972), which had used the method.

The simple structure of this system dynamics can be applied and extended easily by every specialist. The theory of system dynamics is universally applicable because every phenomenon that may be considered as a change in a state can be described by this method and studied by means of simulation. It has already been applied in many disciplines such as biology, chemistry, hydrology, environmental science, ecology, economics, social science and pharmacology. The possibilities of application are determined primarily by the professional knowledge of the researcher and not by his knowledge of mathematics or his ability to use a computer.

This universality will undoubtedly promote an interdisciplinary construction of models, in which systems of one speciality are connected with those of another by paths of information. However, it may be utopistic to think that system dynamics can be used to build and study models in which the real world is elaborated on all levels of complexity. Reality will always be too complex to be modelled wholly in this form. One has to fall back upon the use of aggregated elements, which as we know substantially decreases the degree of generality and applicability.

2 Rate and its resultant; differential equations and integration

The *rate* by which a change in a state of a variable takes place, is expressed in the dimension: amount per time unit. Depending on the nature of the state, this amount can relate to different quantities, such as weight, length, number and rate. The rate itself may be constant for a certain period; it may also change without a clear pattern (at random) or according to certain rules. These so-called *decision* rules must then be converted into differential equations. Note that decision does not have a human connotation; the differential equation describing a chemical reaction can be considered as a decision rule.

It has great advantages to illustrate with simple examples a discussion on nature and function of a differential equation and on the integration associated with this equation. The principles used hereby are essentially the same as for more complex phenomena. The most simple case is the solution of a differential equation to describe the constant speed or rate of change in position of a vehicle. Plotted against time in hours (h) in a graph, this speed (in km h^{-1}) is shown as a straight line parallel to the time axis. What is the result of this speed after a certain period? In other words, what is the distance covered? This question can be answered easily. The speed is multiplied by the length of time or period concerned and the distance covered s in km is obtained as result. One has now integrated the differential equation $\mathrm{d}s/\mathrm{d}t = \mathrm{c}$. Integration is applied chiefly to find the values of the function of a variable when its differential quotient is known.

The differential quotient or derivative $\mathrm{d}s/\mathrm{d}t$ is a notation for the rate during an infinitely small time interval $\mathrm{d}t$. As is usual in a programming language processed by a computer, an abbreviation of the verbal description of the rate concept is often used in the text instead of the differential quotient; for instance, the speed of the flow of water w into a tank $\mathrm{d}w/\mathrm{d}t$ (another notation is $\dot{w}$) may be written as INR and the rate of change in speed (acceleration) as ACC.

In a graph, this integration is the same as the computation of the area bounded by the time axis, by the line parallel to this time axis at a value c of the rate ordinate and by both lines, parallel to the vertical rate axis, at 2 points of time indicating the period. The result of the integration, here the distance covered s in km, can be plotted as a function of time. See the dashed line through the origin with slope c in Fig. 2. In a graph, the slope of a straight line or the tangent to a curve

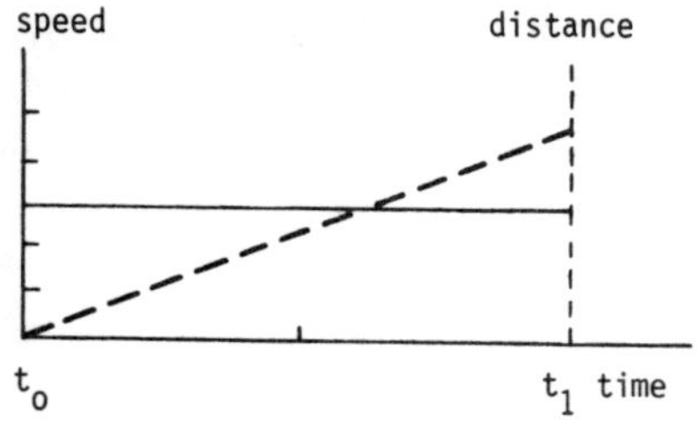

Fig. 2. Speed in km h^{-1} (solid line) and distance covered in km (broken line) as functions of time. The area bounded by the speed line, the two lines at the moment t_0 and t_1, respectively and the time axis equals the distance covered after $t_0 - t_1$ time units.

represents the speed or rate at a certain moment (Fig. 2).

Question 1
In a graph, the distance in meters m on the y axis is plotted against the time in seconds s on the x axis; the result is a straight line.
a. What does the slope of the line represent?
b. What is the dimension of this concept?
c. What can be said about this concept, if the line is parallel to the x axis?

Such a procedure is always performed by integration. The right side of the differential equation is mostly more complex, often so complex that the shape of the integrated function of the variable with respect to time cannot be assessed mathematically. Here one must use the numerical solution. On the other hand, the mathematician always tries to integrate mathematically such differential equations by introducing boundary conditions or constraints into the model. The analytical or mathematical solution often gives a better understanding of the behaviour of the system, but can be applied less easily to practical circumstances.

Question 2
Suppose that the rate of net change in the working capital of a firm, taken as the difference between profit and loss over some years, is given by the graph of Fig. 3.
a. Calculate and sketch the changes in the working capital in the first 6 years, assuming that the initial investment was 6×10^6 guilders.
b. What do you notice?

c. What happens if the scale of the time axis is halved so that the changes in rate do not take place at the time points 2, 4 and 5 but at 4, 8 and 10, respectively (no changes of rate occur after the original point 6)?

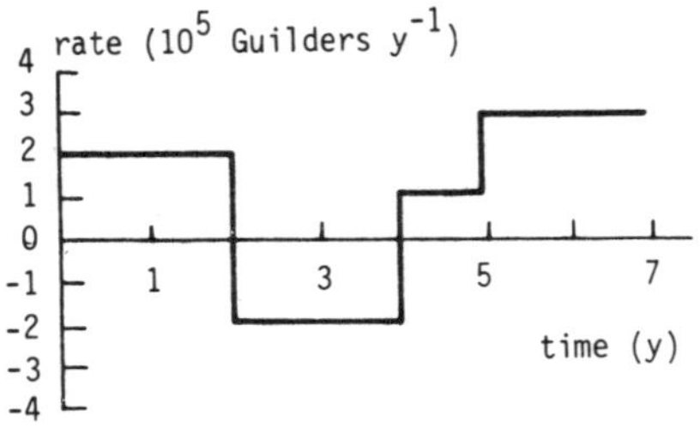

Fig. 3. The rate of change in working capital as a function of time.

The method of direct calculation by determining the area can be used quite well when the rate is a simple function of time. To solve more complex differential equations, this area calculation is also used, but then step-by-step. This procedure will be treated later.

Question 3

In a graph, the rate of water flowing into a tank in litres (l) per second (s) on the y axis are plotted against the time t in seconds (s) on the x axis. The result is a descending straight line with a rate of 1.2 litre per second at point $t=0$ and with a rate equal to zero after 30 seconds.
a. What is the amount of water in the tank after 30 seconds, if there was no water in the tank at point $t=0$?
b. Sketch the amounts of water in the tank as function of time, between the time points $t=0$ and $t=50$.

Often it is possible for a specialist to indicate on which factors and in which way the rate of a process depends. With an increase in knowledge in a science, there is the tendency to make the differential equations more complex so that they can no longer be integrated analytically. Then, if it is impossible to gauge shape and characteristics of the function describing the process with respect to time, the differential equation becomes of little value. This drawback is partly overcome by the numerical solution.

3 Analytical and numerical integration; differential equation and finite-difference equation

Question 3 referred to the problem of filling a tank with water. The constructor could have regulated the valve with a time switch. Then the rate of flow is described as a function of time by the differential equation $dw/dt = -1.2/30 \times t + 1.2$ in which w is the number of litres and t the time in seconds; consequently, dw/dt is the changing rate with which the water is flowing into the tank.

Question 4
How is this equation deduced algebraically from the data given in Question 3?

An analytical integration of this differential equation gives the amount of water w in the tank as a function of time according to $w = -1.2/60 \times t^2 + 1.2 \times t$. With this equation, it is now possible to compute the inflow of water for each period between the points $t=0$ and $t=30$.

Question 5
How can the result of Question 4 be verified?

In this last example, the rate of flow is not dependent on the amount of water already present in the tank. The rate need not always be independent of the state as the next example shows.

Suppose the ecologist thinks, on biological grounds, that the number of animals in an area increases by a certain percentage every year. Then the rate of increase, expressed in numbers of animals per year, would be determined by the number of animals already present and consequently would not be constant in successive years. Under such conditions, the following differential equation holds: $dy/dt = c \times y$, in which y is the number of animals at a certain moment and c the percentage of annual growth. The rate is now a function of the number of animals y; in a graph this relationship is represented by a straight line through the origin.

Question 6

What are the graphs of the rate equations with values of c equal to 10% and 5% per year, respectively?

Such differential equations can be integrated analytically and produce the well-known exponential growth curve $y_t = y_0 + e^{c \times t}$, in which t is the time in the chosen units and e the base of the natural or Napierian logarithms. The subscripts of *y* represent time; consequently, y_t and y_0 are the amounts of moment t and at the start of the calculation, respectively: y_0 is the starting value. With this equation, the state of the variable *y* at every moment can be calculated.

Until now we have investigated the influence of a certain rate equation on a state variable by analytically integrating this equation, after which the state of the variable could be represented as a function of the time, the behaviour. It is also possible to construct this last relationship by computing successively the changes during a number of successive short periods; the rate during such a short period is supposed to be constant. One starts with a certain initial state y_0. By using the rate equation concerned, one can calculate the absolute rate during the next time interval or step Δt and the subsequent change in state during this time interval. The new state again causes a new rate which holds for the next interval Δt, and so on.

Question 7

Why is it necessary to calculate the absolute rates (amounts per time unit) at a number of time points?

The following example explains this procedure. Suppose that the rate at which an amount of water *w* is flowing into the tank through an adjustable valve, is given by the differential equation $dw/dt = \frac{1}{4} \times (16 - w)$. Suppose also that there is no water in the tank at $t = 0$; thus $w_0 = 0$. The rate in $l\,s^{-1}$ by which water is flowing at that moment into the tank equals: $\frac{1}{4} \times (16 - 0)$ or $4\,l\,s^{-1}$. If we take the length of the time interval Δt equal to 2 seconds, then 8 litres water will have flowed into the tank after 2 seconds; *w* becomes 8 l. During the following time interval of 2 seconds, the rate is then: $\frac{1}{4} \times (16 - 8)$ or $2\,l\,s^{-1}$. Therefore, during this time step 4 litres is flowing into the tank, so that the total quantity of water in the tank equals to $8 + 4$ or 12 l. The calculation proceeds as follows:

moment		inflow during the interval (l)	amount of water in tank (l)	difference from the maximum of 16 l (l)	rate of inflow ($l\,s^{-1}$)
0	} Δt		0	16	4
2		8	8	8	2
4					
6					
8					
10					
12					

Question 8

Complete the calculation and plot the amount of water in the tank against time.

a. What do you notice?
b. When is the rate of inflow zero?
c. What happens if an 8 is substituted for a 4 in the fraction $\frac{1}{4}$?
d. Suggest a name for this fraction.

The reader has now performed a numerical integration. As we know, analytical integration can also be applied here. Integrating the differential equation $dw/dt = \frac{1}{4} \times (16 - w)$ gives the equation $w_t = 16 - (16 - 0) \times e^{-t/4}$. Fig. 4 shows the state variable w as a function of time.

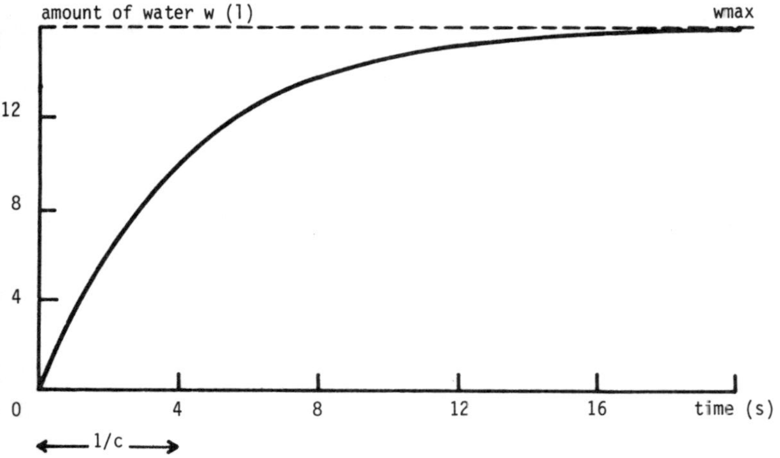

Fig. 4. The amount of water w as function of time. The curve is the integral of the differential equation $dw/dt = \frac{1}{4} \times (16 - w)$ in which w is the amount of water at moment t. The broken line represents the maximum amount wmax = 16 litres.

Question 9

Plot the results of the calculations of Question 8 in Fig. 4.

a. Are the results of these calculations of the amount of water in the tank underestimated or overestimated compared with those of the analytical solution?

b. How do you explain this difference and in which way could it be corrected?

c. If such calculations are also executed with the rate equations of the exponential growth of Question 6, do you expect the numerical solution to underestimate or overestimate?

Actually, with the calculation just discussed a numerical integration is done in the same way as with the computer. The researcher converts continuous differential equations into *finite-difference* equations or *rate* equations, with a *difference quotient* $\Delta y/\Delta t$; with aid of these finite-difference equations the new state or amounts are computed.

The *state* equations describe how the changes are effected and form the integral. Here the multiplication of the rate equation and the length of the time interval Δt are essential. This multiplication can be executed in various ways, the integration method after Euler being the best-known. However, whatever the method a discrete integrated function is always obtained (Kuo, 1965).

The only difficulty is to express the rate and state equations in a form that can be processed easily by a computer. We will refer later to this subject. The use of simulation languages facilitates this part too. These languages include procedures which allow for inaccuracy and errors inherent to numerical integration (see Question 9). However, this book does not go into these matters.

4 Feedback loops

Study of the behaviour of man-made control and servo mechanisms has shown that the structure of a system may be more significant than the individual elements for the system behaviour. An important structure was the *closed loop* or *feedback loop*, in which the state of an element or variable determines the degree of action or flow, which subsequently changes this state. This process takes place in a continuously circulating loop. There are two kinds of feedback loop.

In a *positive* feedback system, the action enhances the state so that afterwards the action becomes greater again. An example is the exponential growth according to $y_t = y_0 \times e^{c \times t}$ with as underlying differential or growth-rate equation $dy/dt = c \times y$. This is the model of, for example, the growth of capital that is put out at a fixed interest a year or of the unlimited growth of algae in a lake with a constant 'relative' or 'intrinsic' growth rate $(dy/dt)/y$, which expression follows from the differential equation. The growth itself as function of time is given in Fig. 5. The absolute increase per time unit is determined by the amounts already present, so that the increase in amounts is enormous. The system of a positive feedback loop produces, as it were, a departure from some reference, neutral condition or goal, often that of zero activity. Such an equilibrium state in a positive feedback loop is often called an 'unstable' equilibrium.

Unlike the positive feedback, the *negative* feedback loop is goal-seeking; a departure from this goal or equilibrium produces an action to return the state to this goal. An example is the mechanism for the automatic filling of a tank with water up to a certain level. The tank is filled according to the equation $w_t = \text{wmax} - (\text{wmax} - w_0) \times e^{-c \times t}$, obtained as we have seen by integration of the differential equation $dw/dt = c \times (\text{wmax} - w)$. The term w_0 is the amount of water in the tank at the beginning of the calculation (see Fig. 4). With the help of the differential equation, trace how this negative feedback loop works.

Question 10

The death rate of a group of animals is described by $dy/dt = -c \times y$, in which y is the number of animals and c the relative death rate.

a. Does this system contain a positive or a negative feedback loop?

b. What is the goal or equilibrium of the system?

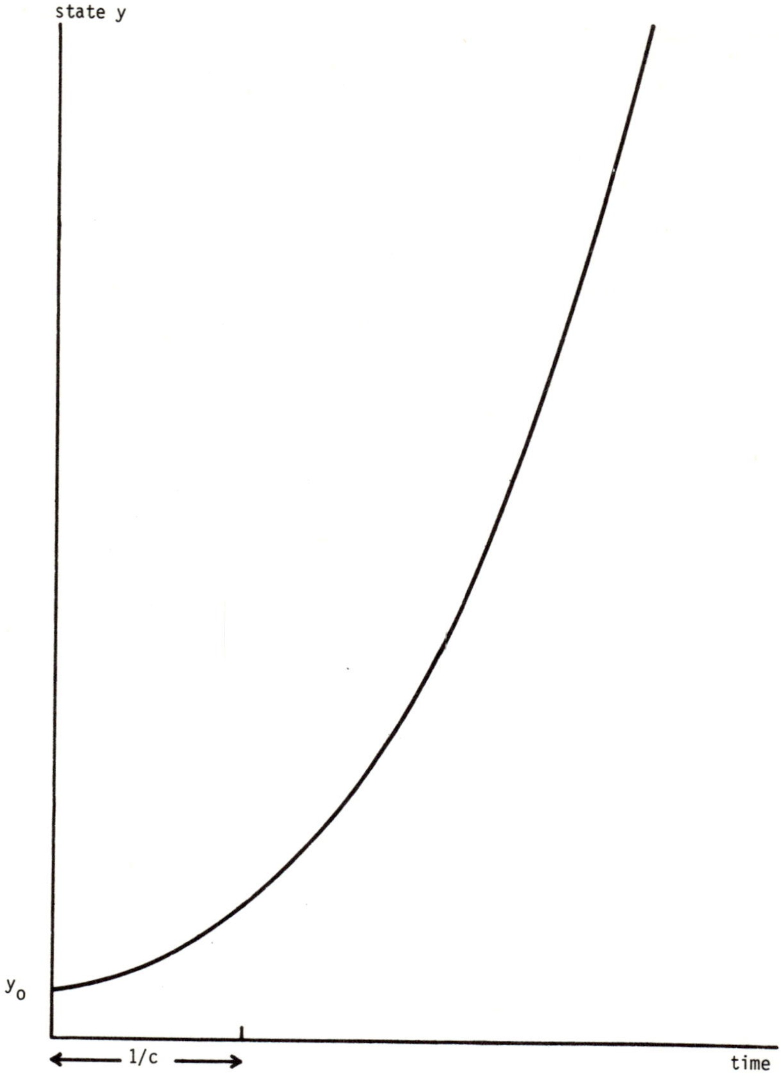

Fig. 5. Graph of the exponential growth curve $y_t = y_0 \times e^{c \times t}$ as integral of the differential equation $dy/dt = c \times y$.

The principle of the feedback structure is present everywhere in the world around us. We see that every decision on an action or flow has to be based on information about the state of one or more elements of the system. Such concepts as decision, action and information should not always be given a human connotation. These concepts include not only processes such as deciding on the length of a rest period in relation to degree of tiredness and the state or interaction between a government's programme and a population's reaction, reflected in how it votes, but also the working of a system describing exponential growth, the reaction of living things to poisonous chemicals and the automatic filling of a tank up to a maximum.

The elementary model of a feedback structure comprises four characteristic elements: a closed boundary, the state, the information on this state and the decision function controlling the action by which the state is altered. Fig. 6 shows schematically the relationships.

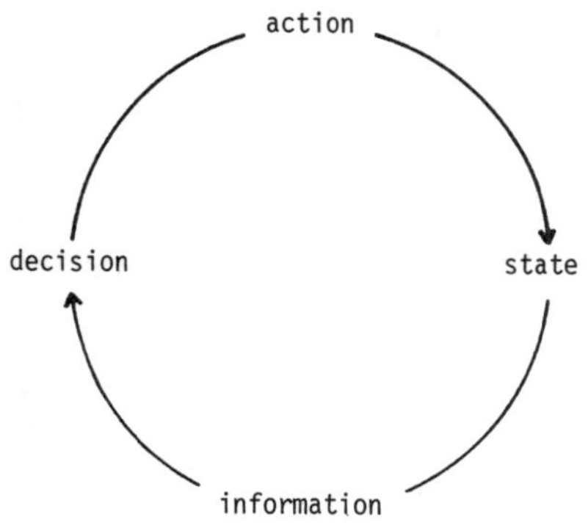

Fig. 6. Scheme of the feedback loop.

Because the system has a closed boundary, its behaviour must be accounted for by the structure only; the behaviour arises from the properties inside the system. Although factors outside the system do influence it, they are not essential for the pattern of behaviour (see the origin of the behaviour given in Fig. 1). Furthermore, the closed whole contains a feedback loop with a decision on the action built in. State and decision functions are parts of this loop, connected by an information chain or flow. The state is an element that is altered by the action or flow; the result is integration. The state contains all information on earlier changes and is, as it were, a memory. Through the decision function, the state determines the action by which it is altered. The components of a decision function are the goal, the state as observed by the decision function, the discrepancy between this state and goal and finally, the necessary action resulting from this discrepancy. The system of feedback can only be described by a function that produces the same mathematical function when integrated and successively differentiated. Discussion of the 'why' will be referred to later, in Chapter 8.

Question 11
Describe the process of buretting or that of steering a bicycle.

The importance of the feedback structure and how it works can be illustrated best by two comparable examples, with or without a feedback structure. Both examples refer to the filling of a tank with water through an adjustable valve and are worked out in Fig. 7 by *relational diagrams.*

In the first example, there is no connection between the water level in the tank and the aperture of the valve; the valve is not 'aware' of the water level and has an aperture that does not change. In the second example, however, the system is 'aware' of the water level in the tank; through the float in the tank and the lever between float and valve, information about the water level is transmitted to the valve; the water

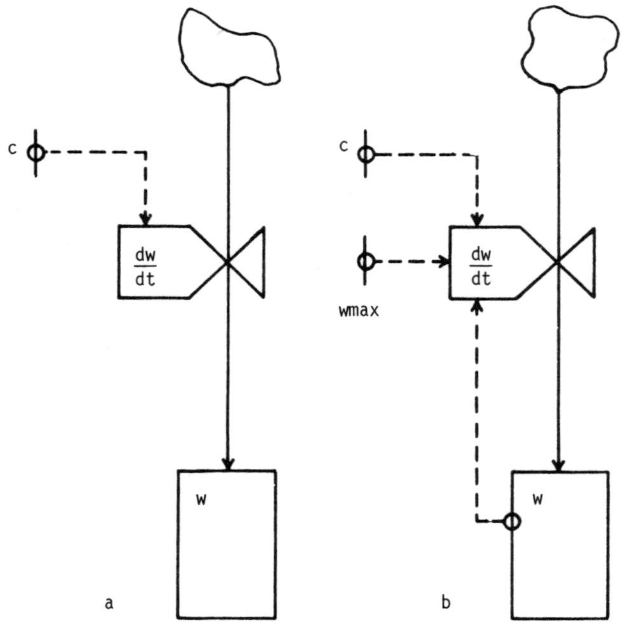

Fig. 7. Relational diagram of a system of the filling of a tank with water, a. without feedback loop and without maximum level, b. with feedback loop to a maximum level. For the explanation of the symbols, see Fig. 9.

level determines the position of the valve and thus the aperture and the flow rate or decision. The builder of this system fitted a valve whose aperture closes when the water level in the tank has attained its maximum. The system reacts instantaneously to information; it is striving for a goal, namely the maximum level. Consequently, this system has a negative feedback loop.

In the first example, the rate of flow is constant and independent of the water level, and the relevant differential equation is $dw/dt = c$. By integration, one can derive that the amount of water w at every moment can be calculated from $w_t = c \times t$. In the example with the feedback structure, the aperture of the valve on which the rate of flow depends, is a function of the water level and therefore not constant. How is this function derived? It is reasonable to suppose (cf. the process of buretting), that the rate of flow is a constant fraction of the difference between the maximum level wmax and the instantaneous level *w*. The lower the water in the tank, the faster the flow and conversely. The differential equation now becomes $dw/dt = c \times (\mathrm{wmax} - w)$, from which after integration the amount of water in the tank can be calculated as function of time according to $w_t = \mathrm{wmax} - (\mathrm{wmax} - w_0) \times e^{-c \times t}$; c is the constant fraction value and w_0 represents as usual the initial state. The rate of flow is zero at the moment that the water level in the tank has reached the maximum (goal). This situation is called an equilibrium because the rate of flow at that level has become zero. This state must be distinguished from a stationary state, steady state or dynamic equilibrium (in German: Fliessgleichgewicht), at which the total amount present is no longer changed, but the rate or rates do not equal zero.

5 Time coefficients

To formulate the two differential or rate equations which represent a positive or a negative feedback system, the rates were calculated by taking a certain percentage of a state or of a difference between a target state or goal and the instantaneous state per time unit. What dimension has this proportionality factor? To answer this question, it is possible to compare the dimensions on both sides of the equals sign in the rate equation; these should equal each other. In fact it is recommendable to always check dimensions when forming equations, especially of complex situations; by following this procedure, one learns more about the significance of the various coefficients and parameters, and can often see how to determine them. Such a check guards against computer difficulties.

In both cases, this comparison shows that the proportionality factor c has the dimension 1/time or time^{-1}. Only then do the dimensions on both sides of the rate equation equal each other. This time element is significant for the behaviour of the system and is called the *time coefficient* TC. In the two systems treated so far, the time coefficients were assumed to be constant. It is also possible that they are not constant or are assumed to be not constant during the process. In the equations for exponential growth or for the system of automatically filling a tank, the time coefficients equal the inverse of the proportionality factor c. The equations of the exponential growth become $dy/dt = 1/TC \times y$ and $y_t = y_0 \times e^{t/TC}$, respectively.

Question 12

What are the time coefficients if the growth percentages are 5, 2 and 0.1% a year, respectively?

The time coefficient determines mainly the reaction rate and indirectly the behaviour of the system. This can be demonstrated by the most simple feedback systems: the exponential growth curve and the system by which a tank is filled automatically with water. In such feedback systems with only one integration, the time coefficient determines the degree of reaction. Check this statement by comparing the integrated growth curves with the time coefficients of Question 12.

Question 13
What can be said about the amount of water which flows into the tank, with a constant absolute rate, after one time coefficient?

In such simple examples, the time coefficient can be defined as the time needed to bring the system into equilibrium, assuming a constant absolute rate during that period; this assumption is expressed by a tangent to the integrated function at that point of time, extending from that point to the intercept with the equilibrium line.

Question 14
a. Check this statement by using the rate equation of the system for automatical filling a tank with water. This property of the time coefficient indeed holds for every point of the integrated function!
b. Such a property also holds in a positive feedback system. However, the formulation is different. Why?

Question 12 and the preceding text gave the relationship between increase or decrease in percentage per time unit on one side and the value of the time coefficient on the other. Theoretically, this rule of thumb is incorrect because for this relationship it was assumed that the absolute rate during a period equal to the time coefficient does not change. However this rate, or geometrically expressed, the tangent to the integrated curve during that period, is changing continuously. The time coefficient calculated according to this assumption is therefore underrated for an exponential increase and overrated for an exponential decrease. The values of these deviations can be assessed easily by comparing the time coefficient calculated by the rule of thumb with that calculated according to the integrated equation $y_t = y_0 \times e^{\pm t/TC}$. For a growth percentage of 10% a year and with an initial state of 100, this equation becomes $110 = 100 \times e^{t/TC}$. By taking logarithms of this equation, it follows that TC does not equal 10 years but 1/ln 1.1 or 10.48 years, i.e. a difference of about 5%. This difference becomes larger at higher percentages. At low values of c, for instance smaller than 10%, the rule given in Question 12 does not give deviations that are too large.

The time coefficient is obviously significant because it, or a derived form, is accentuated by well-known names in the various sciences: the relative or intrinsic growth rate $(dy/dt)/y$ with the dimension $time^{-1}$, the doubling time, the half-life, the time constant or transmission time of the control-system theory, the turnover time, the average total residence time and the delay time.

Question 15
The doubling time, defined as the real time needed to double the amount, equals $0.7 \times TC$ and is therefore smaller than the time for the situation in Question 13.
a. Why?
b. How can the factor 0.7 be derived?
c. What could be the definition of half-life or half-value time? The real half-life period equals $0.7 \times TC$ for the same reason.

The *relaxation time*, often used in physics, is the time needed to decrease the state to 1/eth or 0.37th part of the original value. It is the time coefficient of the exponential return to the original state and can be used as a measure of the speed with which a system is absorbing disturbances.

As discussed in Question 10, the death of individuals in a group is described by the differential equation $dy/dt = -1/TC \times y$ of which the integrated function is $y_t = y_0 \times e^{-1/TC \times t}$. This model is illustrated in Fig. 8. The death rate is mostly given as a certain percentage per time unit, for instance 12.5% a year. The time coefficient is therefore about 8 years. What is the average lifetime or average total residence time of

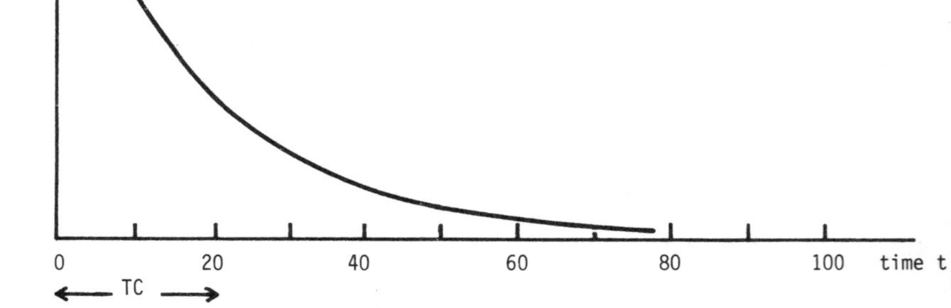

Fig. 8. The exponential death curve $y_t = y_0 \times e^{-t/TC}$ as an integral of the differential equation $dy/dt = -1/TC \times y$. After a period of TC time units the value of 37% of the initial state is reached; this relationship holds for every part of the curve. However, with the assumption that the absolute rate will be constant the equilibrium state will be reached after a period of TC time units.

the individuals of the group? It is common to speak of an average lifetime of 65 years with the implicit assumption that death takes place with a constant percentage. What is the time coefficient in this case?

Suppose that 10 persons enter a room at the same moment and leave this room at different moments: 1 person after 2 hours, 4 after 3 hours, 2 after 4 hours, and 3 after 5 hours. What is the average total residence time of these persons in the room? It is easy to understand that this time equals $(1 \times 2 + 4 \times 3 + 2 \times 4 + 3 \times 5)/10$ or 3.7 hours. The average residence time is determined not only by the residence time of each individual but also by the shape of the frequency-distribution function of the moments at which each person leaves the room.

If the departures of persons or particles from a space is expressed as a percentage of the individuals still present in the space, the exponential decrease curve is at the same time the frequency-distribution curve of the moments at which the persons or the particles are leaving the space. It is possible to prove mathematically, in a way analogous to the calculation procedure just described, that the average residence time and the average departure or transit time of the individuals are equal to the time coefficient TC (Goudriaan, 1973).

Question 16

The amount of poison in a fishing pond decreases by chemical decay at a rate proportional to the amount of poison still present. The proportionality factor is called k.

a. What is the dimension of k?

b. Sketch the cumulative frequency-distribution curve of the residence times of the poisonous particles in the pond as a function of time. Where on the time axis is the average residence time?

Another example is the death of biological individuals. If death takes place at a constant percentage of the present individuals and if the mean lifetime is 65 years, the time coefficient TC equals 65 years.

Question 17

a. What is the assumption about the time coefficient in all these cases?

b. Suppose that the death of a group of individuals must be described in this way. Is the assumption made in the first part of this question reasonable? Why? How is it possible to build a more realistic model?

Closely related to the concept of the average residence time is that of *delay time*. This concept involves an intended change of place or form that is not realized immediately. A transformation of raw material into a product requires time for manufacture, the transfer of oil from the mining area through the pipeline to client takes time when the oil is in the pipeline, the transmission of information requires time as information delay, governmental laws take time before they are implemented, decisions can not be realized instantaneously etc. All these delays can be considered simply as the average residence time in a delay element or box. A delay with an exponential curve can be worked out by integration; the time coefficient used is the average delay time. The reverse applies too: every integration yields a delay! We will return to these exponential delays.

Finally, the time coefficient is important for the length of the time interval Δt (DELT). We have already seen that the drawbacks of numerical solutions can be overcome partly by using small time intervals. However, smaller intervals require much computer time and cost a lot of money. Therefore, the tendency to increase the length of the time interval does raise the question how far this enlargement may proceed without invalidating the prediction.

The slope or tangent to a point of the integrated function and the significance of the errors incurred are determined mainly by the time coefficient. As a rule the length of the interval is taken to be not greater than $\frac{1}{5}$–$\frac{1}{4}$ of the smallest time coefficient of the system. If the interval in the simulation procedure is too large, the system's behaviour will have nothing to do with reality (see also Question 18).

The combined effect of more time coefficients with more integrations inside one feedback loop (see Chapter 8 on the building of larger models) has been described in some cases. In a positive feedback loop, composed of m integrations, the ultimate time coefficient of this loop equals $\sqrt[m]{TC_1 \times TC_2 \times \ldots TC_m}$. The time coefficient of a negative feedback loop with 2 integrations goes by the name of *period* and, after working out, equals $2\pi \times \sqrt{TC_1 \times TC_2}$. Mostly the effect of a combination of time coefficients in a loop cannot be expressed so simply. The influence and the significance of the combined time coefficients have to be studied by simulation.

6 Relational diagrams

Strictly speaking, *relational diagrams*, which visualize the elements and relationships (structure) of a model, are not essential. However many people find them useful for building and elaborating more abstract mathematical models. A relational diagram has the following advantages. At the start of research, it summarizes the most important elements and relationships and helps the researcher maintain an overall picture. Especially when problems are complex, it simplifies the working-out of rate and state equations. Incidentally it makes the content and characteristics of a model easily accessible to others. Finally, a relational diagram improves the comprehensibility of a model so that effects on the system's behaviour and the significance of certain structures (loops) for the behaviour stand out more clearly.

A model of the automatic filling of a tank with water, of which the differential equation is $dw/dt = 1/TC \times (wmax - w)$, is represented by the relational diagram shown in Fig. 7. The representation of this and other differential equations is based on a number of agreements summarized in Fig. 9. These agreements do not need further explanation. It is usually sufficient to represent the states, constants, parameters etc. on which the decision function depends by information lines in the relational diagrams. How the decision function itself is composed, is mostly left out of the diagram. This composition is given by the rate equation, to be discussed later. However, the values of the constants are often mentioned in the diagram; in the example already given: $wmax = 16$ litres and $TC = 4$ seconds. Sometimes during the research, it appears that the factor, supposed to be constant, is variable after all. Then the symbol used has to be replaced by something else, perhaps a table, an auxiliary equation or a connection with an integral. One sometimes indicates with a + or − sign whether the loop concerns positive or negative feedback.

Special emphasis must be given to the information flows, represented by broken lines. With these information chains, systems of different nature can be combined: systems with dissimilar quantities such as men, energy, money and material and those from different disciplines. Also larger systems can be composed in this way. Information is only transmitted and not processed, with one exception mentioned below. Thus information is given, directly or indirectly, only to decision functions and never to states. Information chains therefore

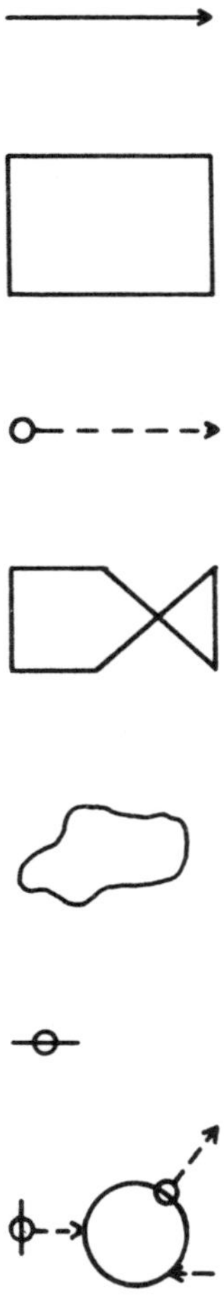

Stream or flow and direction of an action by which amount or state is changed; if necessary, different sorts of line can be used to distinguish between various states but not broken lines.

Integral of the flow; summary of what has happened.

Stream or flow and direction of information; o means that nothing is removed or changed. This is not always so; for instance, if information is delayed, the delay itself is performed by integration.

Valve which indicates that a decision takes place here; the lines of information coming-in indicate on which factors the decision function depends.

Source and sink of quantities in whose content one is not interested.

A constant or a parameter.

Auxiliary equation, which is part of a decision function and is given separately for clarity; conversion coefficients.

Fig. 9. Basic elements of a relational diagram.

connect states to rates. An auxiliary equation is always part of the rate equation and can only be part of an information chain. The information offtake does not affect the information source itself; this source, either a state, constant or variable, is not altered by this information offtake, except when information itself is subjected to a process. This takes place when information is delayed and as such is part of a process; within the process of delaying, changes and integration of amounts of information are executed.

7 Rate equations and state equations

The rate equations and state equations are another representation of the model and specify further the relationships and structures of the model. Connected closely to this is the procedure, including rules and instructions, to calculate the changes during the time interval Δt (DELT). Fig. 10 shows the scheme of the sequence of the calculations.

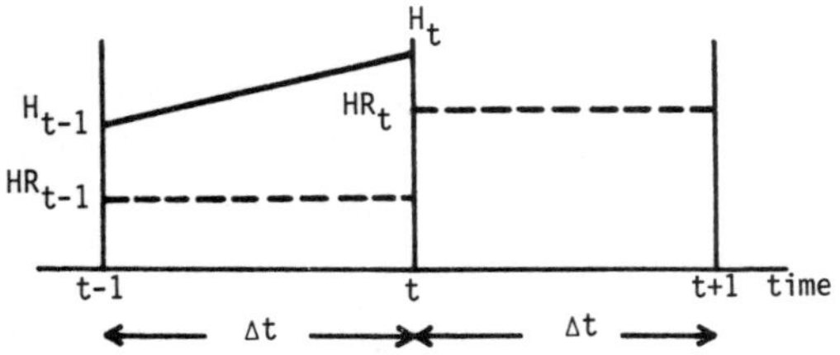

Fig. 10. Scheme of the sequence of the calculations of a state H and a rate HR. Notice that the rate during the calculation interval Δt remains constant, the state, on the contrary, changes.

The vertical in this graph represents 2 axes: the rate axis and the state axis. The rate is represented by a broken line; as the rate is assumed to be constant during the interval Δt, this line runs parallel to the time axis. An agreement is made that every calculation step consists of the calculation of state or quantity H_t at moment t and of the calculation of the succeeding rate HR_t at moment t, using this new state or quantity H_t. This rate HR_t is assumed to be constant during the interval or period t till t+1. For the next calculation step, this time point t and this period are shifted back, as it were, by a time interval Δt; then the state H_t and the rate HR_t become H_{t-1} and HR_{t-1}, respectively. The subscript t of the state means the state exactly at the moment t, whereas the subscript t of the rate indicates the rate during the period $t+\Delta t$ or t till t+1.

When formulating the equations of larger models, it may be important to note this sequence. The initial conditions at the beginning of the calculations must always be considered as belonging to the moment t. Consequently the computing procedure starts with the computation of the rates during the interval 0 till 0+1, which completes the first calculation step and the time subscripts are shifted as described.

The rate and state equations and the necessary (why?) statements about the initial states (INIH etc.) are formulated as follows:

$$H_t = H_{t-1} \pm DELT \times HR_{t-1}$$
$$HR_t = \pm 1/TC \times H_t$$
$$INIH = a\ value$$

The first equation says that the new state or quantity at moment t equals the old one plus or minus the change during the interval Δt. The second equation means that the rate at moment t and during the next interval Δt is a function of a state and of a time coefficient. In the formulation of the ultimate program used by the computer, the time subscripts can usually be dropped. The program itself does not need these subscripts. Besides, the right side of the state equation is formulated differently in a simulation language, for instance in CSMP as INTGRL(INIH, HR), by which the integration as given in the first equation is executed automatically. The use of such simulation languages is not discussed in this book (Shannon, 1975).

Question 18

An interval DELT equal to twice the time coefficient can cause fluctuations. Show this by using the equation for the system of automatic filling of a tank.

The right sides of the rate equations and of the state equations can be extended in various ways. More rates can be taken up in a state equation. The right side of a rate equation may contain every combination of state variables and constants required by the problem. Furthermore, the number of rate equations and state equations can be increased as specified by the content of the problem. See Chapter 8 on the building of larger models.

Question 19

It is known, that the birth rate and the death rate of a population of 50 000 persons are 5% and 2%, respectively.

a. Draw the relational diagram for this system.
b. Formulate the rate and state equations describing this system.

When the equations are formulated, the following points have to be taken into account. The time interval DELT is only found in state equations. It is reasonable to assume that a rate does not depend directly on another rate and because a rate can be determined in reality only indirectly through states rate equations contain only states, other variables and constants, but never rates. The states are altered only by rates. These principles are demonstrated in the relational diagrams by the fact that rates and states alternate with each other.

The dimension of an element in the equation does not determine whether it has a rate function in the system. To check the rate function itself in the system, it is assumed that the system has been stopped; then the real rates become equal to zero, and the other elements remain as information. The presence of a negative feedback loop is given by the minus sign of the state in the rate equation or by the minus sign of the rate in the state equation.

8 Building of larger models

Only the elementary structure of the feedback loop has been treated so far. It is possible to construct models with larger and more complex structures using the principles and agreements mentioned. This enlargement can be obtained by
- increasing the number of integrations,
- using combinations of positive and negative feedback loops,
- introducing non-linear functions,
- constructing larger feedback loops and combining more feedback loops.

An increase in the number of integrations means that the number of state equations, and consequently the number of rate equations, increase too; the order of the differential equation, representing the whole system, increases.

Complexity can be increased by an increase in the number of feedback loops even in structures with only one integration. Each feedback loop may be positive or negative. A combination of positive and negative feedback is present in the model for the growth of yeast, by which alcohol is produced; the alcohol has an adverse effect on the growth rate, the growth even stops at high concentrations. A reasonable assumption is that the relative growth rate (in time^{-1}) decreases linearly with the increase in amount of alcohol in the growth medium. When it is also assumed that the amount of alcohol is a measure of the amount of yeast present, the following differential equation can be formulated: $\mathrm{d}G/\mathrm{d}t = G \times 1/\mathrm{TC} \times (1 - G/\mathrm{GMAX})$, in which G is the amount of yeast at a certain moment and GMAX the maximal amount of yeast. The relational diagram is given in Fig. 11.

Question 20

a. What is assumed about the time coefficient in the diagram?
b. Which feedback loops are present in this system?
c. Describe the effect of each loop in the course of time, using the rate equation.
d. Sketch the amount of yeast as a function of time.
e. Sketch the rate as a function of time.

By increasing the number of integrations the scope of a feedback

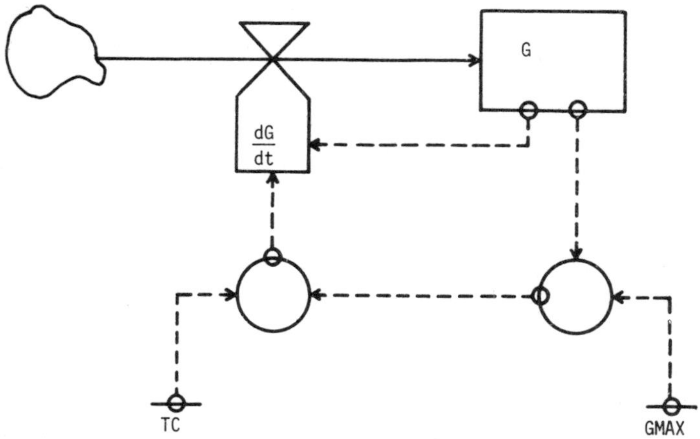

Fig. 11. Relational diagram of the growth of yeast G with the assumption that the time coefficient is not constant during the growth process. The differential equation of the growth rate is: $dG/dt = G \times 1/TC \times (1 - G/GMAX)$.

loop can be extended. Moreover, a state may be part of more feedback loops. Two systems can differ from each other merely by a difference in feedback at the same place in the system.

Non-linearity means that two or more elements somewhere in the system, which change in course of time, are multiplied by each other. An example is the model of logistic growth, which has just been discussed. Here two changing quantities, in this case the same, are multiplied. This non-linearity has a significant influence on the behaviour of systems; for example, it achieves a transition from one mode of behaviour characterized by a positive feedback to another characterized by a negative feedback, and conversely. Non-linearity in a system is also accountable for the well-known insensitivity of changes in behaviour to alterations of the parameter values.

Until now, only simple systems have been discussed. In the application of this system-dynamics approach to problems of a special scientific or management area, the scientist soon meets systems that have to be described by differential equations that cannot be solved analytically. Then the behaviour of such a system can only be studied further by applying a numerical solution with subsequent simulations.

On the whole, there is still little known about the behaviour of larger systems. Increasing the number of elements soon results in feedback loops that are too large for drawing universally applicable conclusions. An example of a slightly more complex but still simple structure is shown by the relational diagram in Fig. 12. The rate equation R1 has the same formulation as that used for the system of the automatic

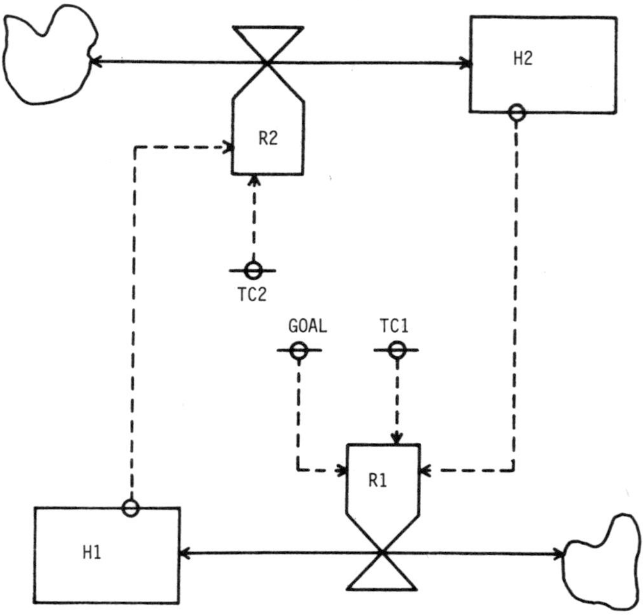

Fig. 12. Relational diagram of a system with a negative feedback loop with two integrations.

filling of a tank. The rate is now influenced by the discrepancy between H2 and its goal.

Question 21

a. What are the rate and state equations in this system with 2 integrations?

b. What kind of feedback loops are present in this system?

The behaviour of this system is characterized by oscillations in H1 and H2 and in R1 and R2, their mathematical formulations being sine and cosine functions (Fig. 13). These oscillations are sustained and not damped. The system reaches the goal or equilibrium but does not stay there. This should occur, when this system with 2 integrations contains another second negative feedback loop. The oscillations of this system are damped and attenuate to the goal. A relational diagram of a system with this behaviour is presented in Fig. 14. This could be the model of a system in which the stock H2, diminished by sale, is replenished by orders. By placing orders H1, of which the number per time unit R1 depends on the discrepancy between the present and

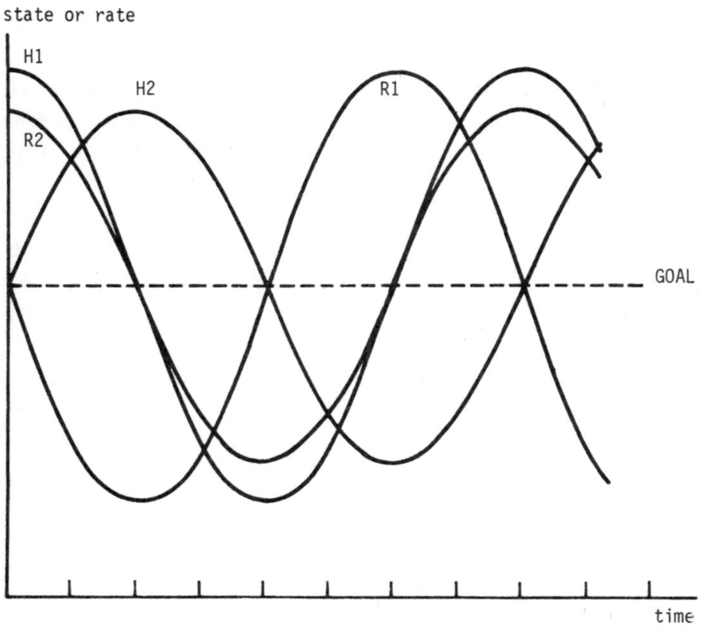

Fig. 13. Behaviour pattern of a system of which the relational diagram is given in Figure 12. Notice the shifts in time of H1, R1, H2 and R2.

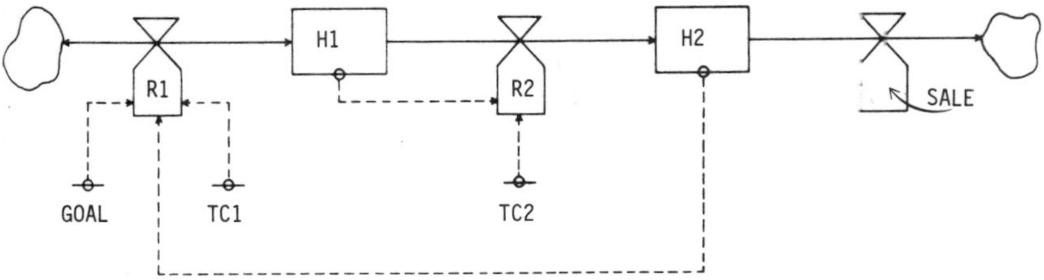

Fig. 14. Relational diagram of a system with two integrations and with two negative feedback loops. The rate SALE is an external factor.

stock or goal that considered necessary and on the time coefficient TC1, an attempt is made to replenish this stock. This replenishment through R2 is not realized instantly; it undergoes a *delay*, the extent of which is determined by the delay time TC2.

Question 22

Formulate the rate and state equations of this system.

It appears from these equations and from the reaction of this system to change that there are 2 negative feedback loops. The larger loop starts, for instance, at H2, runs through the information flow, R1, H1 and R2 back to H2; the small one, starting at H1, runs through the information flow and R2 back to H1. The small feedback loop 'controls' the larger one.

The assessment of the nature of a feedback structure in larger systems can sometimes be difficult. One can use the relational diagram and trace the reaction of a system to a change in a rate or state, introduced somewhere in the loop. If the reaction is such that the system is counteracting this change, the feedback loop is negative. Another method is to count in the rate and state equations the number of multiplications of minus signs in the loop. The feedback is negative when this number is odd.

The fact that the exponential and the sine and cosine functions play an important part in the formulation of feedback systems, is related to the following. In a feedback loop, the integration of a decision function is always followed by the differentiation of the state, resulting again in the decision function and so on. This alternating process within a loop can be continued only, if both differentiation and integration yield the same function; only the functions mentioned meet this requirement.

A negative feedback system of the second order generates oscillations because integration of the differential equation of this system yields an exponential function with the imaginary unit i in the exponent. Such exponentials can be resolved into a sum of the trigonometrical functions of sine and cosine.

9 Simulation of delay and dispersion

The concept of delay time and the fact that every integration yields a delay have been mentioned in Chapter 5. Delay means that an increase in a rate like input is not realized instantly in the output but later and according to a characteristic pattern or frequency-distribution curve. With one integration, this distribution is exponential and equals the curve describing the automatic filling of a tank with water; this change in the output of the delay is realized only slowly. It is possible to simulate some characteristic patterns of delay-response by a number of successive integrations.

Suppose that many ships are leaving a harbour at the same moment and that the average transit time (or residence time or delay time) is 14 days. Only a few ships will take exactly 14 days to reach the destination because there are slow ships and fast ships. Hence the arrival dates show a dispersion according to a certain frequency-distribution curve. These dispersions are met in all kind of problems. Some examples are: the difference in dates of seed germination, the difference in biological response times to signals or to manipulation, the difference in physiological development of biological subjects, the dispersion during the transport of chemical by water in rivers and in soil. Finally, a delay is sometimes found whereby a change in input results in a change in output, delayed indeed but not dispersed, the so-called pipeline effect.

These dispersion-distribution curves can be obtained by simulation, using a cascade of successive integrations. An example is the relational diagram of Fig. 15, which represents a second-order exponential delay

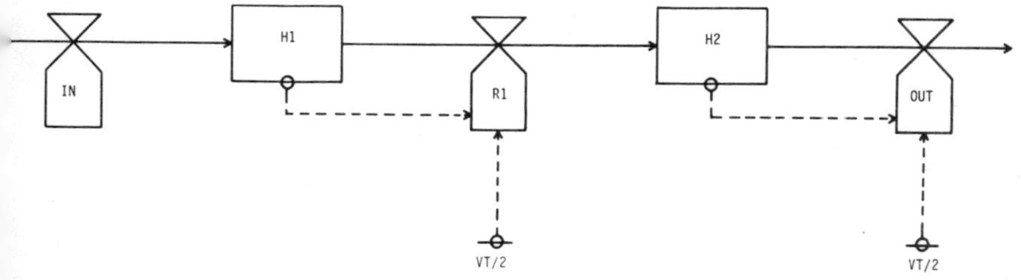

Fig. 15. Relational diagram of an exponential delay of the second order of a rate. The input IN is changed; this change is effectuated in the output in a delayed and transformed way.

of a rate. It is a delay of the second order (why?) with a total delay time VT. The corresponding rate and state equations are:

$$H1_t = H1_{t-1} + DELT \times (IN_{t-1} - R1_{t-1})$$
$$R1_t = H1_t/(VT/2)$$
$$H2_t = H2_{t-1} + DELT \times (R1_{t-1} - OUT_{t-1})$$
$$OUT_t = H2_t/(VT/2)$$

Question 23

a. Why has VT/2 been taken as time coefficient in both rate equations?

b. What are the initial conditions of H1 and H2, assuming that a steady state is reached for a constant inflow rate IN?

Delays of higher order than the second order can be formulated in a similar way. Using simulation languages, one does not need to formulate these equations; they are present as functions of the simulation language used. Delays in information flow can be described and used in the same way.

We have already met the exponential dispersion curve of a delay of the first order. It appears that cascades of successive first-order delays yield dispersion curves distinct from this one. Some of these curves are brought together into Fig. 16. In this graph, the simulated response of the output on a sudden but permanent change in the input is given. This figure shows that the greater the order of the delay, the steeper the distribution curve and the narrower the distribution. The relation-

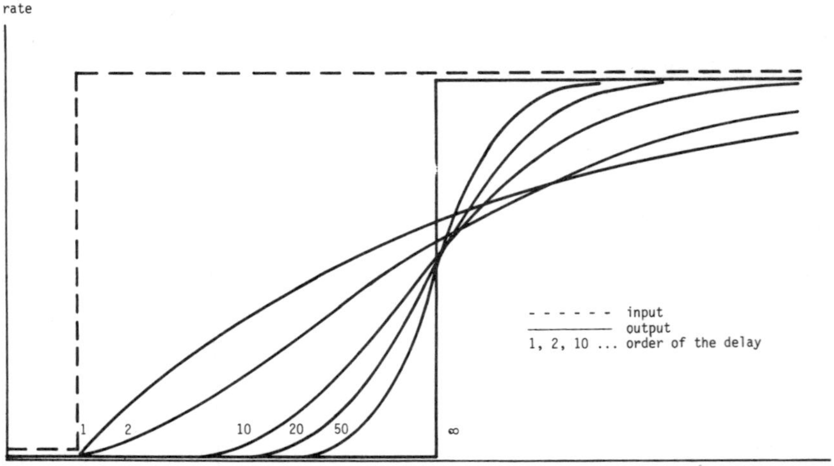

Fig. 16. Patterns of responses of the output rates on sudden changes in the input rates.

ship is formulated by the expression $N = (VT/s)^2$ in which N represents the order or the number of integrations, VT total delay time and s the standard deviation in time units (Goudriaan, 1973). In a delay of infinite order with N equal to infinite, this dispersion disappears and the pipeline effect is obtained. In the model the formula can be used to simulate the empirically found dispersion with a certain delay time.

Question 24

Sketch the output of a sudden but permanent decrease of the input, subject to exponential delays of the first, second, 10th, 20th, 50th and infinite order.

Integrations or delays of a momentary sudden change of a rate, the so-called impulse, cause mostly a transformation in the change too. It is possible to show by reasoning comparable with that used in Exercise 25, that the output of an impulse, subject to a delay of the first order, gives the well-known decrease as a function of time after an initial sudden increase; however, even after passage of the delay time the output is still responding. These initial and subsequent reactions are determined by the ratio of impulse to delay time.

An impulse delay of higher order yields an asymmetrical bell-shaped distribution curve of the output; the output does not react instantly to the impulse. Afterwards the reaction increases strongly and passes after a maximum into a decrease and disappears ultimately. Also in these cases, the output is still reacting after the delay time. At increasing orders of exponential delay, the positions of maximal reaction are shifted to later moments together with a decrease in dispersion; by a delay of infinite order, the output develops into the pipeline effect, i.e. an output of the impulse without any dispersion, delayed with the total delay time.

This artificial dispersion is inherent to the use of exponential changes in the model. However there are many real phenomena without any dispersion and the introduction of exponential changes into the model yields undesirable effects and is not permitted. A well-known example is the age-class in demography; in conformity with the definition, the content of an age-class is shifted to the next one after, for example, each year, without any dispersion. It is possible but impractical (why?) to simulate such shiftings by an exponential delay of infinite order with a delay time of one year. In the literature (de Wit & Goudriaan, 1978), a number of boxcar-train techniques are described, by which a more practical shifting without any dispersion is obtained. One always has to investigate which description is needed and is correct.

10 Exercises

By answering of questions, the reader had been able to test whether he has understood the text or not. By doing the following exercises, he can gain experience of formulating and using system dynamics models. At the same time these exercises provide the opportunity to study some behaviour patterns mentioned already in the text. Hence the working out of these exercises is an indispensable part of this study book.

The exercises are given in order of increasing complexity and perhaps are progressively more difficult. They cannot be solved properly until the whole text has been read through at least once. Not only the solutions should be studied; it is essential that the reader struggles with the problems and finds the answers out himself.

The first three exercises relate to the graphic representation of an integration; some attention is also paid to dimensions. Subsequently, some simple exercises (4–11) with one state or one integration are given. The reader should gain experience of the formulation and comprehension of rate and state equations and improve his insight into the significance of the time coefficient for the behaviour of the most simple systems with positive and negative feedback loops. Exercises 12–21 are more complex and introduce models with more integrations. Then the effect of combining both kinds of feedback loop inside the system are introduced, experience being gained from reading differential equations. Exercises 24 and 25 refer only to the behaviour patterns as affected by exponential delays.

Exercise 1

In a graph the speed (km $hour^{-1}$) on the y axis is plotted against time (hour) on the x axis; this yields a straight line parallel to the x axis.

a. What can be said about the speed?

b. What represents the area between this straight line and the x axis, measured between two instants?

Exercise 2

The growth rates (kg dry matter ha^{-1} day^{-1} during the growing season of a perennial crop are known. These rates are plotted against time yielding the following graph (only the data needed are given).

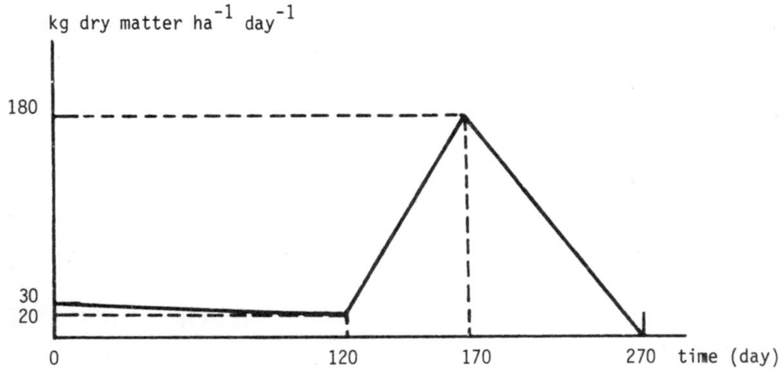

Note. The notation 'kg dry matter ha^{-1} day^{-1}' is preferable to that of 'kg dry matter/ha/day', as the latter is ambiguous; it can mean 'kg dry matter day ha^{-1}' and 'kg dry matter ha^{-1} day^{-1}' the latter being intended.

a. Calculate with the graphical data the amount of dry matter per ha present after 270 days of growth.

b. Sketch and discuss point by point the development of this amount of dry matter per ha during that period.

Exercise 3

1. The management of a flower auction market is interested in the supply (number of cut flowers per week). This supply can be calculated with the aid of an auxiliary equation, in which this supply SUP at a certain moment can be expressed as a function of the number of glass-houses GLA, the number of plants per glass-house PLAGLA and the mean number of flowers FPR by SUP = GLA × PLAGLA × FPR.

What is the dimension of the number of flowers FPR?

2. In the literature on the effect of advertizing charges, the following equation is often used to describe the normal loss of customers and the increase of customers by advertizing as a function of time: $\mathrm{d}S/\mathrm{d}t = \mathrm{r} \times \mathrm{A} \times (\mathrm{M} - S)/\mathrm{M} - \mathrm{q} \times \mathrm{S}$ in which

t = time in months
S = number of customers at moment t
A = advertizing charges in guilders a week
M = saturation point of the market, consequently the maximal value

of S
r and q are constants.
a. Which processes represent each part of the right side of this equation, separated by the minus sign?
b. What is the dimension of r; suggest a name for this constant in connection with this advertizing problem?

Exercise 4

Radioactive material RAM decays with a rate DECR which depends on the amount of material present: a certain ratio of this material per day disappears. We have 1 milligram of this material of which the relative decay rate per day or disintegration constant equals 0.05.
a. Draw a relational diagram of this system.
b. Sketch a graph in which the decay rate (mg a day) is plotted against amount of material present (mg).
c. To which equilibrium value does this system tend?
d. What value has the time coefficient of this system?
e. Write out the equation for the rate of decay.
f. Calculate the decay rates and the amounts of material during the first 30 days as function of time, using a calculation time interval of 5 days.

Exercise 5

A number of systems are formulated by the following rate and state equations and initial values.

1. $L_t = L_{t-1} + \text{DELT} \times R_{t-1}$
$R_t = 2 \times L_t / 4$
$\text{INIL} = 4$
$\text{DELT} = 4.$

a. Draw the relational diagram.
b. What is the value of the time coefficient?
c. Which feedback loop is present in this system?
d. What is the equilibrium value?
e. Is the value of the time interval well-chosen?

2. $L_t = L_{t-1} + \text{DELT} \times R_{t-1}$
$\text{INIL} = 20$
$R_t = (10 - L_t)/6.$

a. Draw the relational diagram.

b. Which feedback loop does this system have?
c. What is the equilibrium value?
d. How large is the time coefficient?
e. What is the rate at moment $t=0$?
f. Which value can DELT have?
g. Sketch the development of L as a function of time.

3. $L_t = L_{t-1} + DELT \times (R1_{t-1} - R2_{t-1})$
$INIL = 2$
$R1_t = 0$
$R2_t = (1 - L_t)/20.$

a. Draw the relational diagram.
b. Which feedback loop is present in this system?
c. What is the equilibrium value?
d. Sketch the development of L as a function of time.
e. How does L develop, if the initial value INIL is -2?
f. What is L's value after one time coefficient? This question is to be answered without the numerical integration technique.

Exercise 6

A system is represented by the following equations:

$HW_t = HW_{t-1} + DELT \times HWR_{t-1}$
$INIHW = 0$
$HWR_t = 1/INS \times (MAH - HW_t).$

a. Which feedback loop does this system govern?
b. With a value of INS to 4 seconds the system shows a certain behaviour. What happens if INS equals 8 instead of 4?

Exercise 7

The relative decomposition rate DCOR of a herbicide or weedkiller HERB in the soil amounts to 0.7, whilst 50 kg per ha of this herbicide is given every year as a single application.
a. Draw the relational diagram of the behaviour of the herbicide in the soil.
b. Which feedback loop does this system display?
c. Discuss and sketch the change in amount of herbicide in the soil in course of time.

Exercise 8

The rate with which photosynthesis reacts to changes in light and carbon dioxide can be formulated by the equation $dP/dt = 1/TAU \times (P_s - P)$, in which P is the rate of photosynthesis in $kg\,CO_2\,m^{-2}\,s^{-1}$ at moment t, P_s the stationary photosynthesis rate belonging to certain light flux and carbon dioxide densities and TAU the time coefficient. The light flux density fluctuates and is alternatively high or low for a while; the changes set in suddenly. The carbon dioxide density is constant.

a. Which process is represented by the differential equation?

b. Sketch the changes in the rates of photosynthesis P and P_s, generated by these sudden changes.

Exercise 9

The beet cyst nematode NEM multiplies only by growing of beets. The annual relative growth rate of a nematode population INCR is constant and equals 1.0. When beets are not grown, the annual relative fall-off rate DEATR equals 0.25.

a. Draw the relational diagram for increase and decrease in nematodes.

b. Form the rate and state equations of this diagram, if the initial number of nematodes is 1000 per ha.

c. Sketch the changes in the numbers of nematodes during 10 years, assuming a crop rotation with beets once in 5 years.

Exercise 10

From experience, the gamekeeper of an area thinks that a roe population ROE of 400 is optimum. He tries to keep this number stationary by issuing shooting permits; thus the number of roes that the hunters are allowed to shoot annually and indeed that are killed, is limited. The problem is to derive how many permits R3 the gamekeeper can issue annually to maintain the initial population of 400. The birth rate of the roes R1 is 12.5% per year, the natural death rate R2 10% per year. Shooting takes place throughout the year.

a. Draw the relational diagram.

b. Write the rate and state equations and the values of the constants.

c. How many permits must be issued annually?

d. Is the model described realistic? What will happen if the initial population is greater or less than 400?

Exercise 11

In a crop-rotation system, the organic matter content ORMAT of the plough layer is enriched by stubble, roots and green manure with an annual rate R1 of 4000 kg per ha; this enrichment is distributed uniformly over the year. It is also known that on the average the annual relative decomposition rate R2 of the organic matter in the soil is 0.02 and that 0.5% of this organic matter is leached annually from the plough layer. The agricultural consultant likes to know which level the organic matter content of the plough layer will reach ultimately.

a. Draw a relational diagram of supply and loss of the organic matter in the plough layer.

b. Give the rate and state equations and the values of the constants.

c. Calculate with the aid of the available data, the organic matter content of the plough layer for which the stationary or steady state is reached.

Exercise 12

The differential equation of the motion under gravity or of the free fall are: $\mathrm{d}v/\mathrm{d}t = \mathrm{g}$ and $\mathrm{d}s/\mathrm{d}t = v$, in which t is the time, g the gravitational constant, v the speed and s the distance covered, all in the well-known units.

a. Draw the relational diagram of the system of free fall.

b. Formulate the rate and state equations.

c. Which feedback loops does this system have?

Exercise 13

How long can the population POP of a country still utilize the current stock of 5000 million tons of coal COL? At this moment, the size of the population is 5 million people, whilst the birth rate BIRTR is 100 000 a year. It may be assumed that the relative birth rate will remain constant; the average life expectancy after birth is taken as 65 years.

a. Draw the relational diagram for the changes in population and coal stock.

b. Derive the rate and state equations, with the assumption that the average consumption of coal is 1 ton year^{-1} person^{-1}.

Exercise 14

A factory wants to establish a pension fund for its employees when they retire. The management sets up an inquiry into the increase and decrease of this fund by contributions of the employees, interest from the capital and by payments to retired employees. The calculations are started with an initial capital of f 1 million given by the factory, with an initial number of employees of 300 and without any pensioners. The interest which is always added to the capital is 5% a year. The contributions of the employees to the fund are fixed at f 500 per employee per year. It may be assumed that the number of employees appointed every year is 10% of the number of employees already present and that 8% of the employees retire every year. The average death rate of the pensioners is 20% a year.

a. Draw a relational diagram of the employees EMP, of the capital volume CAP and of the number of pensioners PENS.

b. Give the initial state, the values of the constants and the rate and state equations.

Exercise 15

The frequency of the annual births in a country is 1 to 20 adults. After 6 years on the average, these children are sent to school. This schooling continues for 10 years on the average, after which everybody is considered to be adult. The mean life expectancy after that is still 50 years. Suppose that the number of babies, of children at school and of adults are 20, 3000 and 100 000, respectively at the beginning of the calculations.

a. Draw the relational diagram for calculation of the changes in the numbers of babies BAB, of children at school SCHO and of adults ADUL.

b. Write the initial conditions, values of the constants and rate and state equations concerned.

c. Which feedback loops does this system contain?

Exercise 16

A town with a specified number of inhabitants has been projected in one of the newly reclaimed polders in the Netherlands. The rate with which the houses are completed and occupied immediately is proportional to the discrepancy between the maximum number of houses projected HOUMAX and the number of completed and occupied house HOU. The houses are built without central heating, but after some time the occupants tend to install it. The central heating dealer of the new polder wants to plan his work and supposes that the rate with which central heating systems are ordered is proportional to the discrepancy between the number of houses occupied and the number

of houses with central heating CH (delivered).
a. Draw the relational diagram for the calculation of the changes in the number of houses occupied and in the number of houses with central heating.
b. Write the rate and state equations concerned.
c. Sketch the curves representing the number of houses HOU and the number of houses with central heating as functions of time.

Exercise 17

The growth rate of algae AGR is a function of the amount of algae by $AGR_t = RAGR \times A_t$, in which A_t is the amount of algae at moment t expressed as the nutrient N in $g\,m^{-3}$ taken up and RAGR is the annual relative growth rate. Moreover, RAGR itself is a function of N by $RAGR_t = 1/TC \times N_t$ in which N_t is expressed as $g\,m^{-3}$. The quantity of N in the nutrient solution is decreased by the uptake and is not replenished. Compare this situation with that of Question 20.
a. Draw the relational diagram of this system.
b. Formulate the rate and state equations.
c. What is the dimension of TC?
d. Which feedback loops does this system contain?
e. Sketch and discuss the changes in N, A and the growth rate AGR as functions of time. Hint: translate the amounts of nutrient into amounts of A.

Exercise 18

In an investigation on the influence of phosphate on growth, plants are grown in pots. The relative growth rate RGR in grams of dry matter per time unit and per gram of dry matter already present has an empirically found maximum RGRMAX. The RGR decreases rectilinearly with increasing plant weight G (in gram). This plant weight too has an empirically found maximum GMAX at which growth stops.
a. Draw the relationship between RGR and G; complete this graph with the characteristic quantities given.
b. From a derive the equation of RGR as function of G.
c. What is the equation for plant growth as function of weight?
d. Draw the relational diagram of this system.
e. Discuss and sketch the increase of weight G as a function of time.

Exercise 19

In literature on business economics, one can read that the number of owners of a product often shows an S-shaped curve with respect to time. However, a satisfactory explanation of this phenomenon is not given. A good explanation could be obtained by assuming that owners OWN usually stimulate non-owners by verbal and visual contact. Thus the rate of increase in owners IOWNR is a function of the number of owners. It may be assumed that the number of stimuli per owner per time unit STITC is not changed in the course of time. The effect of these stimuli EFF expressed as the number of new owners per stimulus decreases as the number of owners increases and becomes zero when the number of owners attains the maximum number of owners OWMAX or the saturation point of the market. The decrease itself is rectilinear.

a. Draw the relational diagram of the model.

b. Does this model give an explanation of the S-shaped curve of the number of owners in the course of time? Explain your answer.

Exercise 20

The relative rate of increase of the leaf area, infected by a fungal population, INFA is proportional to the percentage of the leaf area not yet infected. The total leaf area A itself at a certain moment, that means infected plus uninfected, increases in the course of time according to the S-shaped or logistic curve; the maximum attainable leaf area MA is assumed to be constant.

a. Draw the relational diagram of both growth processes combined.

b. Formulate the rate and state equations.

Exercise 21

The Gompertz growth curve is derived from the rate equation $dW/dt = M_0 \times W \times e^{-S \times t}$, in which W is the amount of dry matter in kg, M_0 a constant, S the senility factor and t the time. This senility factor S may be investigated more closely by considering the part $M_0 \times e^{-S \times t}$ as the integration of another rate equation. Then the rate equation mentioned first can be separated into two rate equations of the first order.

a. Formulate these rate equations.

b. Draw the relational diagram.

c. Sketch the development of W as function of time.

Exercise 22

A lake with a water supply equal to amount of water discharged is polluted by organic matter which is decomposed in the lake. On the assumption of a complete mixing, the increase and decrease in amount of organic matter can be described by the following differential equation: $dc/dt = W(t)/V - c/t_0 - k \times c$, in which

c = the concentration of organic matter in the water,
$t_0 = Q/V$,
Q = the supply of water per time unit,
V = the volume of the lake and
W(t) = the amount of organic matter deposited at moment t.

a. Which processes do the 3 parts of the right side of the equation represent?
b. How long on the average will a water particle stay in the lake?

Exercise 23

The equations of Streeter–Phelps describe the changes in amounts of dissolved oxygen O in a river, effectuated by oxygen depletion due to decomposition of organic wastes L and by oxygen replenishments due to atmospheric reaeration. These equations are $dL/dt = -k_d \times L$ and $dO/dt = k_r \times (O_s - O) - k_d \times L$, in which

L = the amount of organic matter in the water, expressed in terms of the biological oxygen demand in mg/l,
O = the amount of dissolved oxygen in mg/l,
O_s = the maximum oxygen content of water at specific conditions or at saturation point,
k_d = the decomposition coefficient and
k_r = the reaeration coefficient.

a. Draw the relational diagram.
b. Give the rate and state equations.
c. Discuss and sketch the changes in the oxygen deficit (the so-called oxygen sag curve) after the introduction of a momentary pollutional load, assuming a prompt complete mixing.

Exercise 24

Arabian countries export oil by tanker to the Netherlands. The rate of transport is 0.2 million ton per week, the mean transit time from export harbour to destination is 9 weeks.
a. Draw the relational diagram of this transport system, assuming an exponential delay of the third order.
b. Formulate the rate and state equations concerned, give the initial

states for a steady state situation.

c. Suppose that the Arabian countries suddenly start a boycott and decrease the oil export to the Netherlands by 50%. Sketch the oil import in tons a week as a function of time, from just before the boycott until 4 months afterwards.

Exercise 25

A production process consists of a number of successive steps: raw material A is converted to B, the product B is converted into C. The processes take place simultaneously. The rates ABR and BCR with which every product is produced is proportional to the remaining quantity of material or product from which it is converted. It is assumed that the relative production rate of ABR is greater than that of C BCR. The production has been started without any quantity of B and C; during the process, no replenishment of A takes place.

a. Draw the relational diagram of both processes.

b. Give the rate and state equations concerned.

c. Discuss and sketch the shape of the curves representing the amounts of A, B and C in the course of time.

Exercise 26

The two banking firms B and M show a remarkable similarity of behaviour to the opening of new bank offices of the respective firm; an explosive increase as it were! It may be expected that arguments for such growth can be found in mutual competition. An explanation could be the following.

The rates INBR and INMR with which B and M, respectively, invest money in new offices, increases proportionally to the total amount already invested in bank offices by the other banking firm; the proportionality factors of both firms are constant with respect to time but may differ from each other. The annual percentage of depreciation with which both firms write off (=loss of money) the formerly paid investments in offices may also differ from each other, but are constant in the course of time.

a. Draw the relational diagram to calculate the changes in amounts of money MB and MM, invested in offices by firms B and M, respectively.

b. Write the state and rate equations concerned.

c. Which feedback loops are present in this system?

d. Derive under which conditions (relationships between the time coefficients) a steady state of the invested amounts will appear in the long run.

Exercise 27

A system is represented by the following rate and state equations:

$$S1_t = S1_{t-1} + DELT \times R1_{t-1}$$
$$INIS1 = 10$$
$$S2_t = S2_{t-1} + DELT \times R2_{t-1}$$
$$INIS2 = 0$$
$$R1_t = -\tfrac{1}{5} \times S2_t$$
$$R2_t = \tfrac{1}{8} \times S1_t.$$

a. Draw the relational diagram of this system.
b. Which feedback loops can be indicated?
c. Calculate the quantities S1 and S2 and the rates R1 and R2 at 10 successive time intervals with a length of 2 time units. What do you notice? Give a simplified example of a system, process etc., that can be described by these 2 equations.

Exercise 28

The following equations hold for the description of the motion of a comparatively long pendulum:

$$a = g/l \times (-P)$$
$$dv/dt = a$$
$$dP/dt = v$$

in which l is the length of the pendulum, t the time, P the discrepancy between pendulum end and the equilibrium position on which $P = 0$, g the constant of gravity and v the speed, all in the well-known units.
a. Draw the relational diagram of this system of the pendulum motion.
b. Write the rate and state equations concerned.
c. Which feedback loops does this system have? Describe the behaviour of this system.

Exercise 29

The interaction between a host-insect H and its parasite-insect P can be described on biological grounds by the two following differential equations: $dH/dt = a \times H - b \times P \times H$ and $dP/dt = c \times P - e \times P^2/H$. The parameters a and c represent the net-influences of the natural birth and death processes; consequently, b and e are the parameters of the additional death processes influenced by the interaction of the two insect species.
a. Draw the relational diagram of this system.
b. Formulate the rate and state equations.

c. Will an equilibrium or a steady state be attained in the long run? If so, at which values of H and P?
d. Which feedback loops does this system contain? Which behaviour pattern can be derived from this knowledge?

Exercise 30

A company for management of road restaurants RORE with 100 small and large locations in Europe has the following management policy: 2.5% of the total annual turnover of all restaurants RORE is used for maintenance and for building of new restaurants; the mean expenses of maintenance per restaurant per year MAICF are estimated at f 100 000, the mean cost of building a new restaurant BUICF at $2\frac{1}{2}$ million guilders. Therefore, the number of restaurants to be built is determined by the amount left when the expenses of maintenance for all restaurants have been subtracted from the 2.5% of the total turnover. The average life of a restaurant DETC is taken to be 20 years and the annual turnover per restaurant TUO decreases with an increasing number of restaurants according to the following graph. On the average, 3 years will pass from commissioning till realization of a restaurant (exponential delay of the third order with a total delay time VT of 3 years). The company is so large that all activities can be considered to be executed continuously.

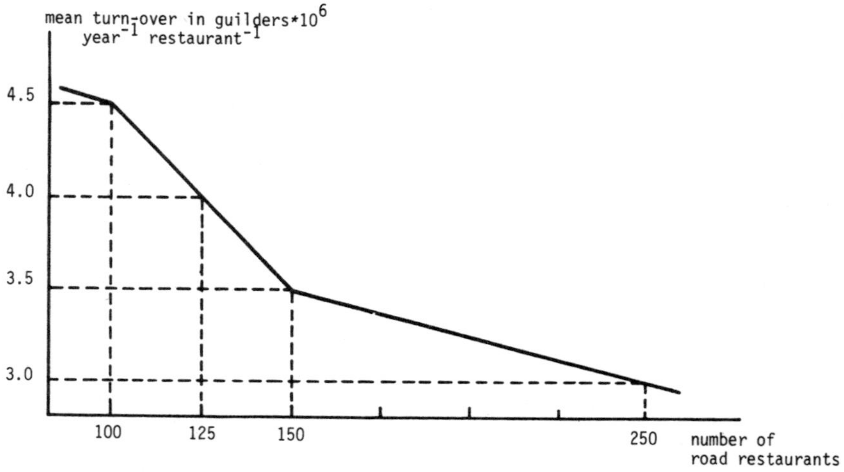

a. Draw the relational diagram of this system, so that the number of restaurants can be calculated as a function of time. Put in the diagram a delay of the first order only.

b. Formulate the rate and state equations; the equations of the delay do not have to be given.
c. Give the dimensions of the separate parts in the rate equation.
d. Is the number of commissions for building restaurants subject to a negative or positive feedback loop?
e. Will a steady state appear in the long run? How many are there if this state is attained?

11 Answers to the questions

Question 1

a. Speed or rate.
b. Meter $second^{-1}$ or meter/second.
c. Speed equals zero.

Question 2

a.

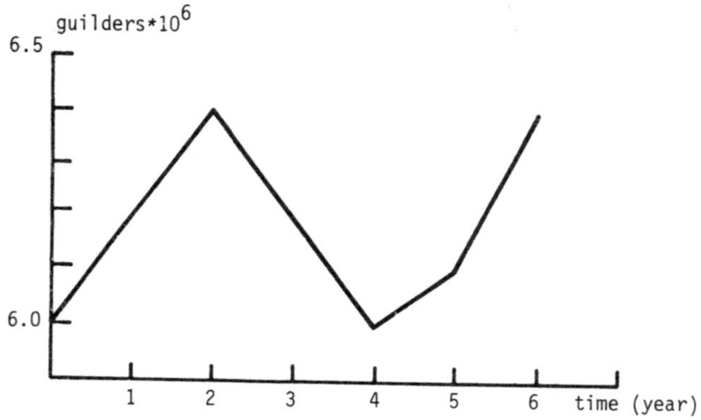

b. There are fluctuations, the maxima and the minima being delayed with respect to the beginning of an alteration.
c. The ampiltude of the fluctuations becomes larger.

Question 3

a. The amount of water in the tank is 18 litres which is equal to the area bounded by the rate of flow line and by both axes.
b. At $t=0$ there is no water in the tank; after $t=30$ the flow stops and the maximum level is reached. The curve lying between these moments can be obtained by calculating the inflow of water during a number of successive time intervals and by summing these amounts; every amount is found by multiplying the average rate during an interval by the length of the corresponding time interval.

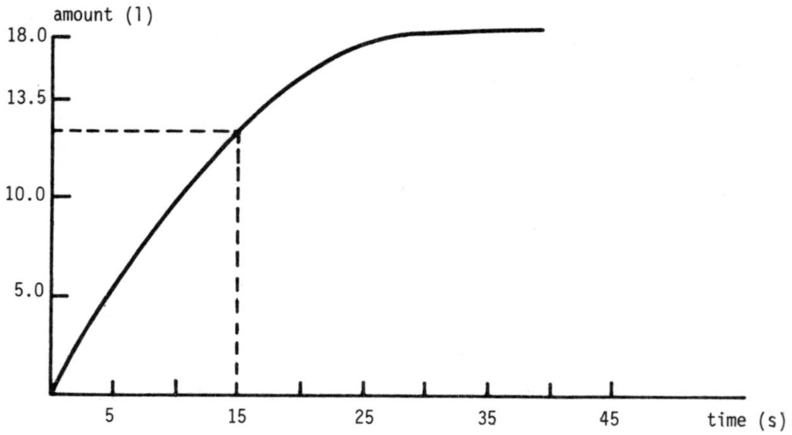

Question 4
The formula of a descending straight line is $y = -a \times x + b$, in which a is the slope of the line and b the intercept with the y axis. Here the values of these parameters are 1.2/30 and 1.2, respectively.

Question 5
Substitute various values for t into the equation of the amount of water in the tank as function of time; for $t = 30$ the substitution results in $w = -1.2/60 \times 900 + 1.2 \times 30$ etc.

Question 6

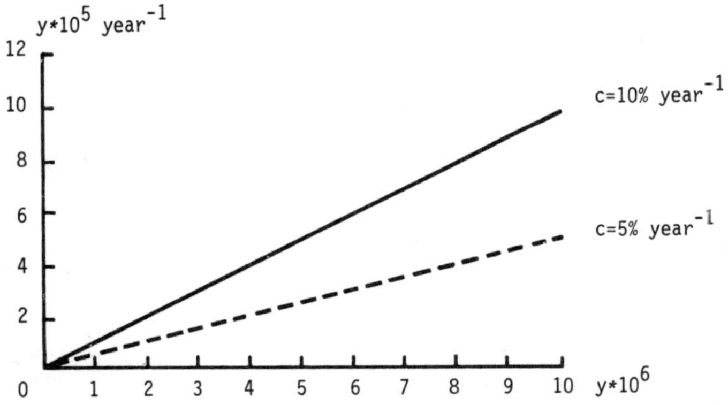

Question 7
According to the equation, the absolute state determines the value of the next change.

Question 8

a. The calculated values do not form a smoothed curve, but a broken one.
b. The rate of inflow is zero when the maximum level is reached, i.e. the tank is full.
c. With an 8 instead of a 4, the process is slower.
d. The inflow coefficient, for instance.

Question 9

a. The result of the calculations gives overestimates compared with those of the analytical solution.
b. This overestimate is caused by the wrong assumption that the rate is constant during the calculation interval. It could be corrected by making the interval smaller.
c. Underestimate.

Question 10

a. A negative feedback loop.
b. The goal of the process is the zero state.

Question 11

The information from the eye about the level of liquid in the glass allows a decision on the change in position of the glass tap, which causes a decrease in the rate of outflow at the tip. By intuition, the size of the opening of the tap will probably be proportional to the observed discrepancy between level and desired level or goal. This procedure also holds for the steering of a bicycle.

Question 12

The time coefficients are 20, 50 and 1000 years, respectively.

Question 13

The amount is duplicated.

Question 14

a. The rate equation is given by $dw/dt = 1/TC \times (16 - w)$; this relationship holds for every moment. The change during a time interval equal to TC and with constant rate equals $TC \times 1/TC \times (16 - w)$ or $(16 - w)$. Thus the difference between w and the equilibrium value of 16 is bridged within one interval equal to the time coefficient.
b. In a positive feedback loop, the change is directed away from the equilibrium state, so that the direction of the extrapolation of the

tangent must be reversed toward the unstable equilibrium state; the tangent cuts the horizontal equilibrium line after a time-coefficient interval.

Question 15

a. As the true state is increasing continuously the actual duplication is attained earlier.

b. The increase in the amount during the time interval Δt equals to $e^{\Delta t/TC}$ according to the exponential growth equation; for a duplication this expression for the increment must equal 2. Hence $\Delta t = TC \times \ln 2$ or $\Delta t = 0.7 \times TC$.

c. If the process is decreasing the amount, the half-life is the time necessary to reach half the amount.

Question 16

a. Time^{-1}

b. The cumulative frequency-distribution curve of the residence times (and of the transit times) equals the descending exponential curve, reversed with respect to a line parallel to the time axis.

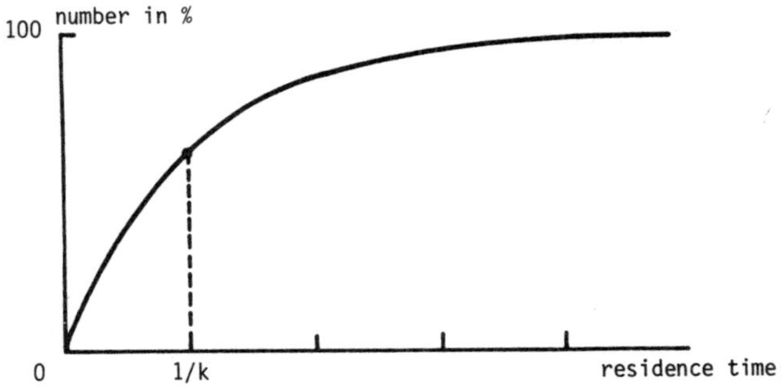

Question 17

a. It is assumed that the time coefficient continues to be constant.

b. No; the death risk increases with age and therefore, the actual time coefficient must decrease. A more realistic model is obtained by taking more age-classes, each with their own time coefficient.

Question 18

The equations are:

$$HW_t = HW_{t-1} + DELT \times INR_{t-1}$$
$$INR_t = \tfrac{1}{4} \times (16 - HW_t).$$

Take a time interval DELT equal to 8; then, 32 litres flow into the tank during the first interval. In the second interval, the inflow is negative (=outflow) because the rate becomes negative according to the rate equation. After the second interval, 32 litres flowed out etc.

Question 19

a.

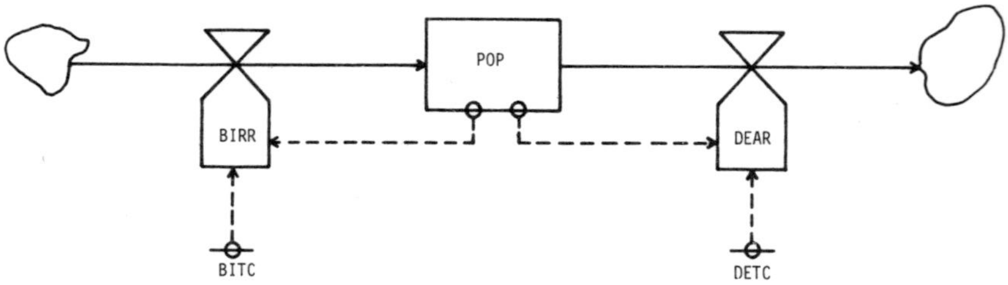

If the process is described only as a net increase or net decrease by birth and death, it suffices to use one flow or stream and the related function in the diagram.

b. $POP_t = POP_{t-1} + DELT \times (BIRR_{t-1} - DEAR_{t-1})$

INIPOP = 50 000 persons

BITC = 20 years

DETC = 50 years

$BIRR_t = \frac{1}{20} \times POP_t$

$DEAR_t = \frac{1}{50} \times POP_t$.

Question 20

a. The time coefficient is not constant and depends on the state of the system.

b. A positive feedback loop from G through information to dG/dt, whose action affects again G; a negative feedback loop through GMAX and G to dG/dt etc.

c. At the beginning of the growth with G small compared with GMAX, the part $(1 - G/GMAX)$ is about 1; in this situation, the positive feedback loop in the part $1/TC \times G$ dominates. In later phases, G is large compared with GMAX and the part in brackets, so the negative feedback loop becomes progessively dominating.

d.

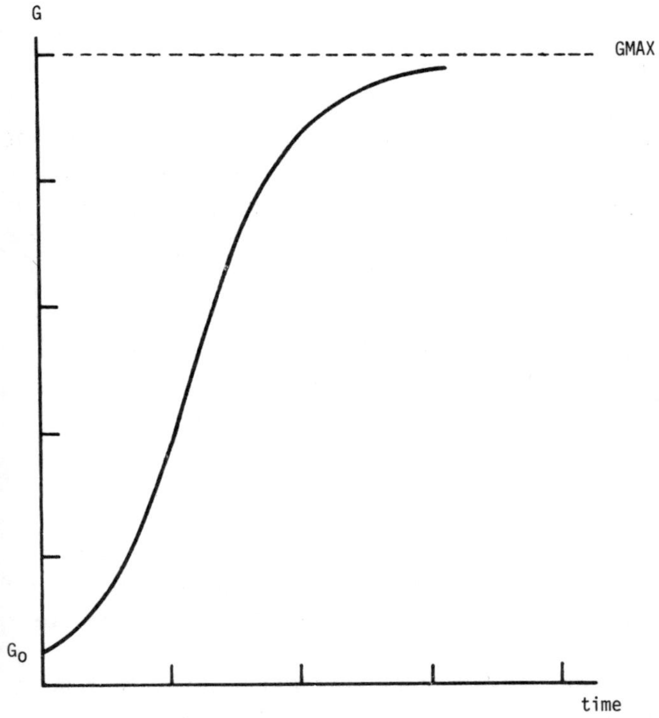

e.

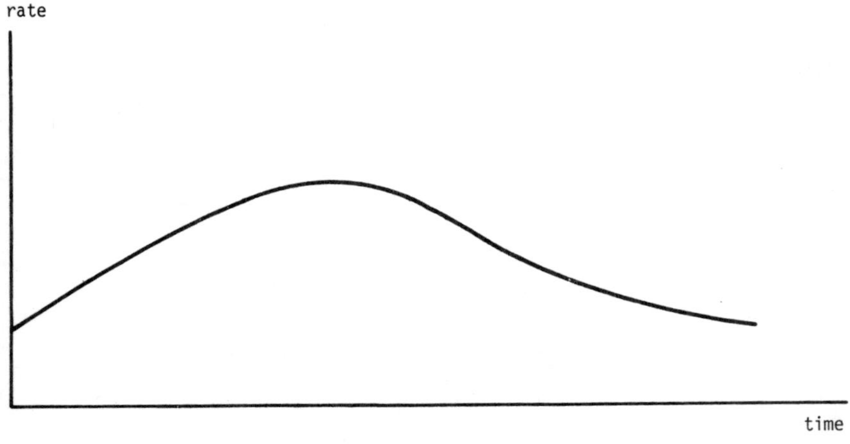

The moment at which the maximum of the rate appears, coincides with the point of inflexion on the logistic curve.

Question 21

a. $H1_t = H1_{t-1} + DELT \times R1_{t-1}$
$R1_t = 1/TC1 \times (GOAL - H2_t)$
$H2_t = H2_{t-1} + DELT \times R2_{t-1}$
$R2_t = 1/TC2 \times H1_t$.

b. The equation shows that, for instance, an increase in H1 will effectuate an increase in H2 through R2. This increase in H2 causes a decrease of the rate R1 according to the rate equation, by which the increase in H1 is undone. Consequently, the model has a negative feedback system.

Question 22

$H1_t = H1_{t-1} + DELT \times (R1_{t-1} - R2_{t-1})$
$H2_t = H2_{t-1} + DELT \times R2_{t-1}$
$R1_t = 1/TC1 \times (GOAL - H2_t)$
$R2_t = 1/TC2 \times H1_t$.

Question 23

a. The mean total residence time in the complete delay is VT time units. There is no difference between both delay elements or boxes, the amounts remain therefore on the average VT/2 time units in every delay element.

b. In a steady state, the amount streaming into and out of both delay elements is the same. Consequently, $SN1_t$ and OUT_t equal IN_t. According to the rate equations, the conditions $IN_t = H1_t/(VT/2)$ and $IN_t = H2_t/(VT/2)$ apply, from which $H1_t = IN_t \times VT/2$ and $H2_t = IN_t \times VT/2$ follow.

Question 24

By a permanent reduction in rate of flow a dispersion pattern develops that equals the reverse of the patterns of Fig. 16.

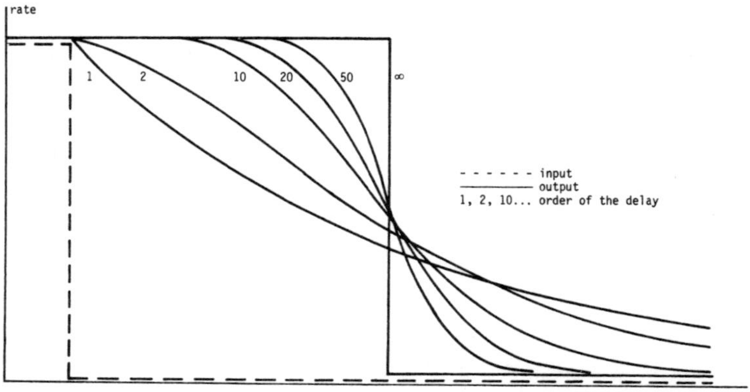

12 Solutions to the exercises

Exercise 1

a. The speed is constant.

b. The distance covered between these 2 months.

Exercise 2

a. After 270 days of growth an amount of dry matter per ha is present which equals the area bounded by the time axis, the growth rate axis and the growth rate lines. Consequently, this amount equals $\frac{1}{2}\times(30+20)\times120+\frac{1}{2}\times(20+180)\times50+\frac{1}{2}\times180\times100$ or 170 00 kilograms dry matter ha^{-1}.

b. The growth rate is decreasing up to 120 days, from 120 till 170 days the rate increases and after 170 days the rate decreases again to zero. The growth curve is given in the following graph:

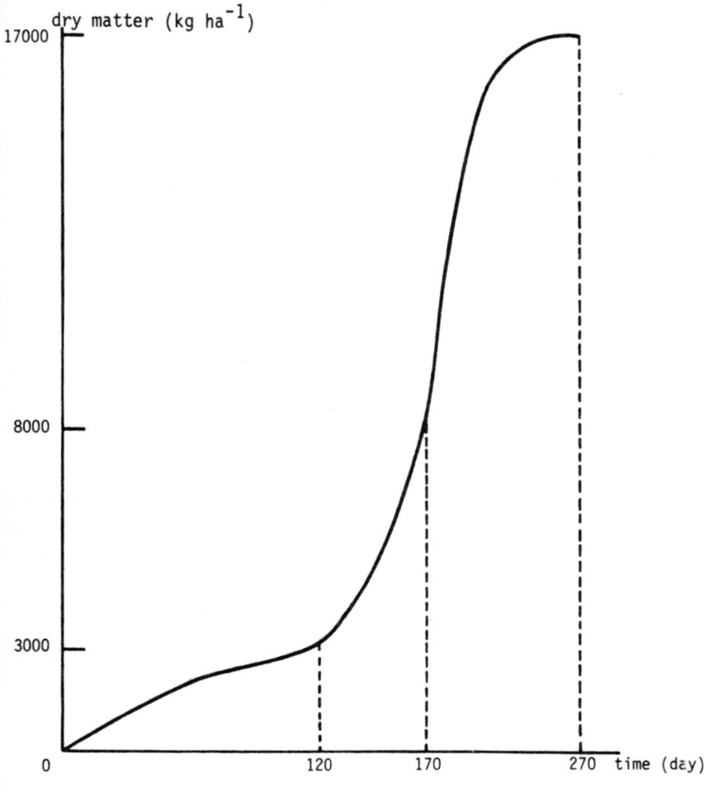

Exercise 3

1. By writing out the dimensions of the left and right sides of the equation, the dimension of the flowering FPR becomes: flower plant^{-1} week^{-1}.

2a. The first part represents the rate of increase by advertizement, the second part describes the normal loss of customers with time.

2b. The dimension of r is: customer guilder^{-1}. A name could be: advertizement efficiency.

Exercise 4

a.

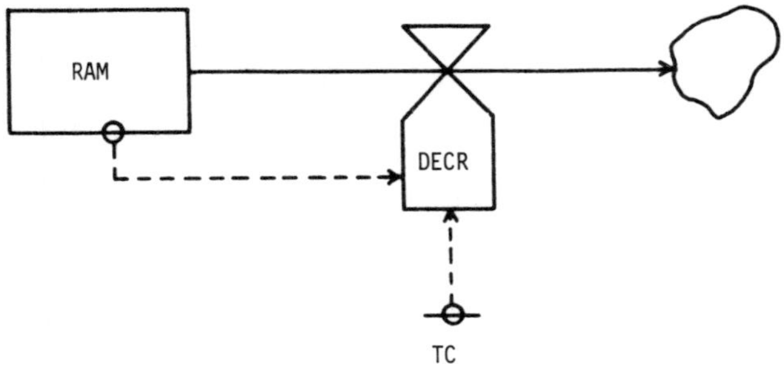

b.

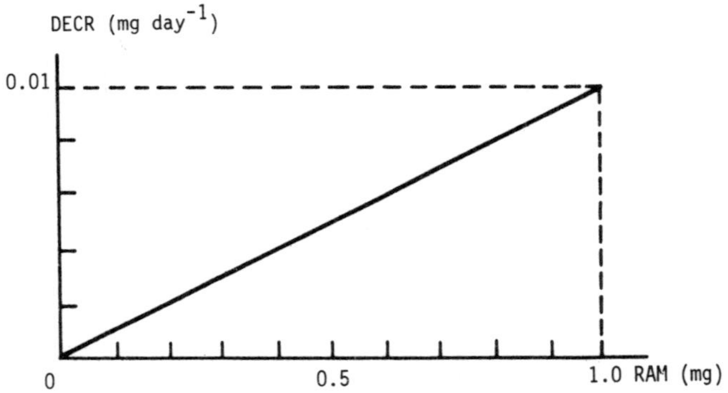

c. At the equilibrium value $dRAM/dt = 0$, thus $DECR = 0$. It follows that $0.05 \times RAM = 0$ or $RAM = 0$.

d. $TC = 1/0.05$ or 20 days.

e. $DECR_t = \frac{1}{20} \times RAM_t$.

f.

t	RAM_t	$DECR_t$
0	1.0000	0.0500
5	0.7500	0.0375
10	0.5625	0.0281
15	0.4220	0.0211
20	0.3165	0.0158
25	0.2375	0.0119
30	0.1780	0.0089

Exercise 5

1a.

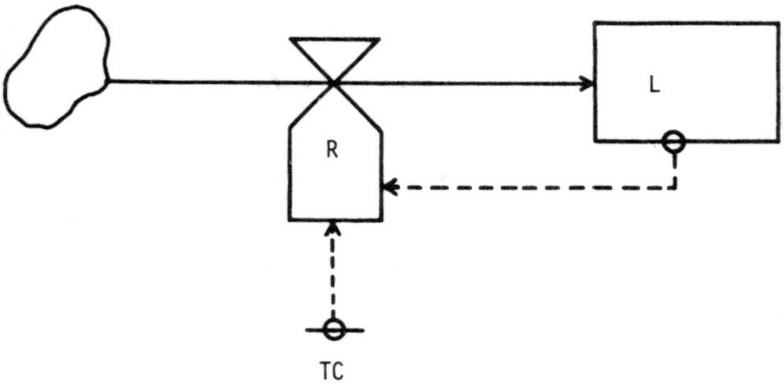

1b. $R_t = 2 \times L_t/4$ or $R_t = \frac{1}{2} \times L_t$, consequently $TC = 2$.
1c. Positive feedback loop.
1d. An unstable equilibrium equal to zero; each discrepancy with this equilibrium causes an exponential growth.
1e. No. Experience has shown that DELT must be about $\frac{1}{4}$ to $\frac{1}{5}$th of TC; by this rule of thumb, DELT can therefore be 1.

2a.

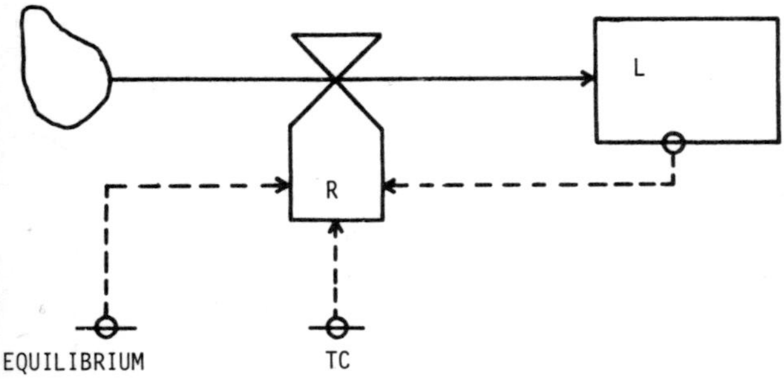

2b. A negative feedback loop by a combination of + and −.

2c. The equilibrium will be reached, when $R_t = 0$ or $(10 - L_t)/6 = 0$, thus $L_t = 10$.

2d. $R_t = 1/TC \times (10 - L_t)$, thus $TC = 6$.

2e. L equals 20 at the moment $t = 0$, thus $R_0 = (10 - 20)/6$ or $-10/6$.

2f. DELT might be 1.5 or 2.

2g.

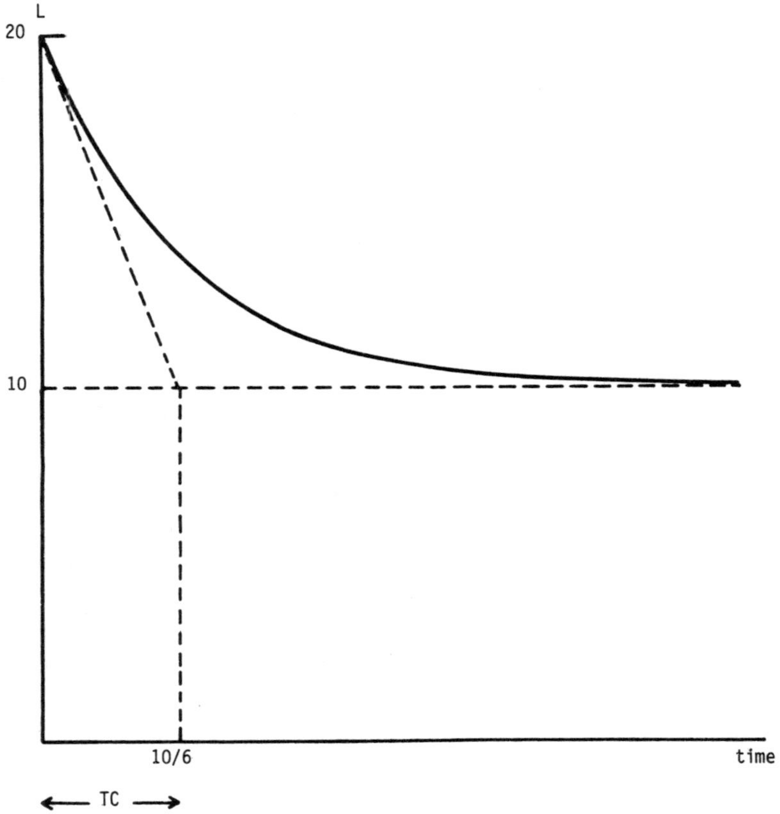

3a.

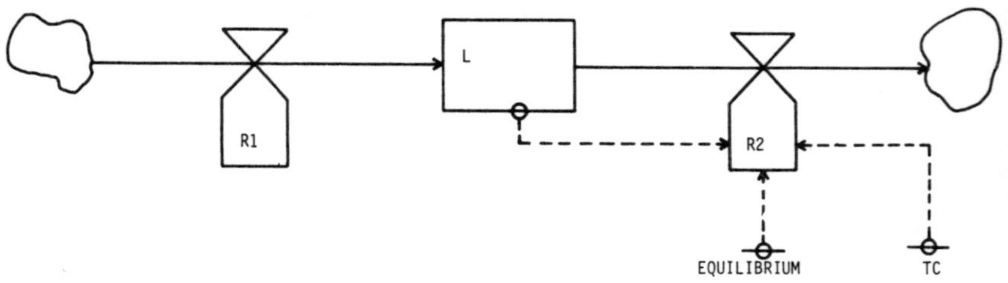

3b. A positive feedback loop by the combination of a negative rate in the state equation and the negative sign of L in the rate equation.
3c. At the equilibrium state $R2_t = 0$ applies, thus by substitution $(1 - L_t)/20 = 0$, which means that the equilibrium value of L equals 1.
3d.

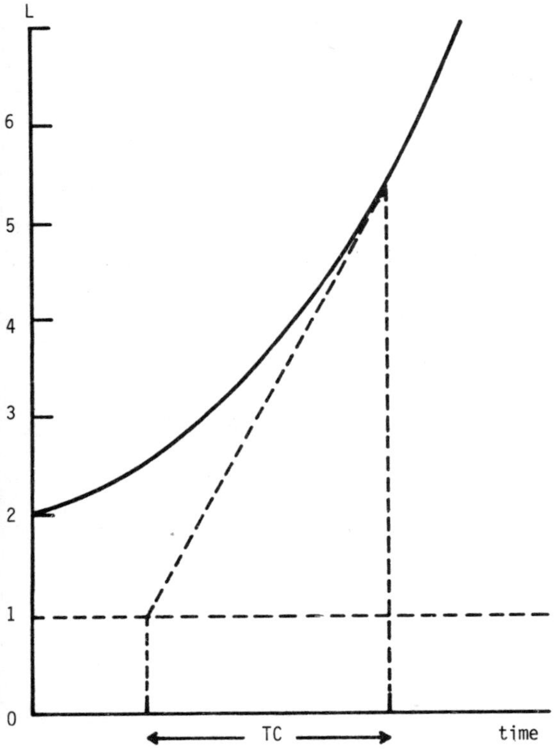

3e. The reverse in respect to the horizontal equilibrium line, the first one shifted slightly to the right.
3f. After one time coefficient or 20 time units the discrepancy of L to the equilibrium value is increased with a factor e. Therefore, after 20 time units $L = 1 \times 2.718 + 1$, in which the first 1 represents the starting value, expressed as difference with the equilibrium value, and the second 1 the equilibrium value itself.

Exercise 6
a. A negative feedback loop by the negative sign of the state in the rate equation.
b. The process state moves more slowly to its equilibrium value.

Exercise 7

a.

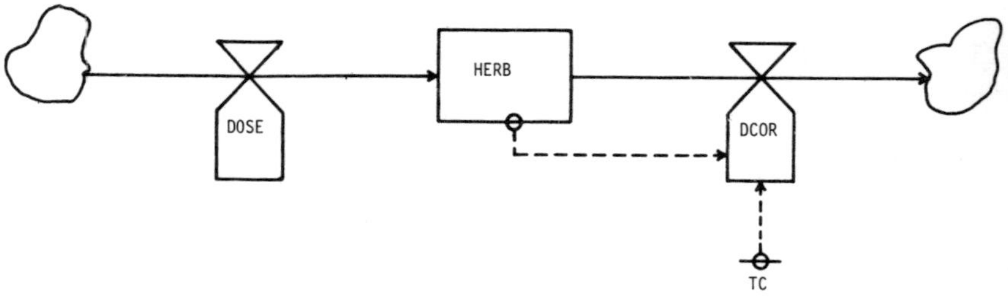

b. A negative feedback loop.

c. The time coefficient is 10/7 years. It is known further, that the half-life is 7/10th of the time coefficient. It follows that 50% of the herbicide in the soil is decomposed after 1 year at every turn so that the amount of herbicide changes as shown in the next graph. The amount of herbicide in the soil will never exceed 100 kg per ha.

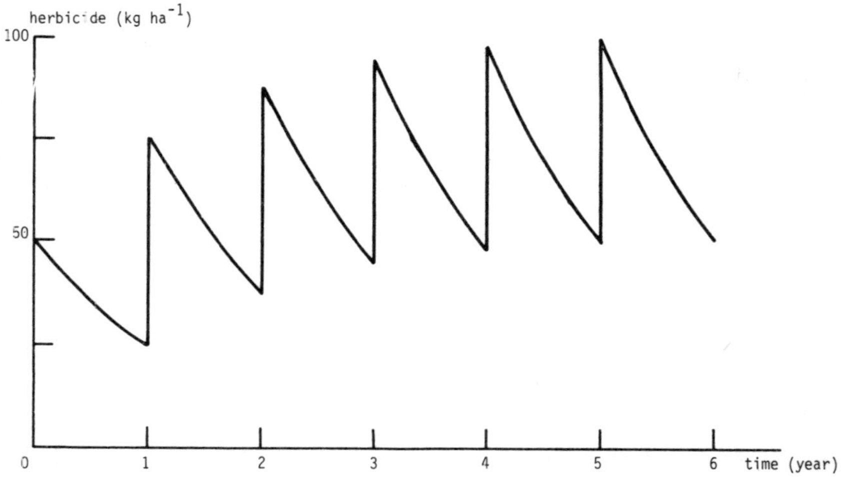

Exercise 8

a. A specific photosynthesis rate P_s is connected with every light flux and carbon dioxide density. Therefore, a change in the latter effects a change in P_s. From the rate equation given, we see that the change in P_s is not reached instantly; the adjustment is exponential.

b. A response is generated as shown in the next figure. To work out this graph, it is assumed that TAU is not affected by the light flux and carbon dioxide densities and is therefore constant. The broken line in the graph is the adjustment value P_s of the photosynthesis rate, which belongs to a certain light flux or carbon dioxide density. The solid line represents the actual photosynthesis rate.

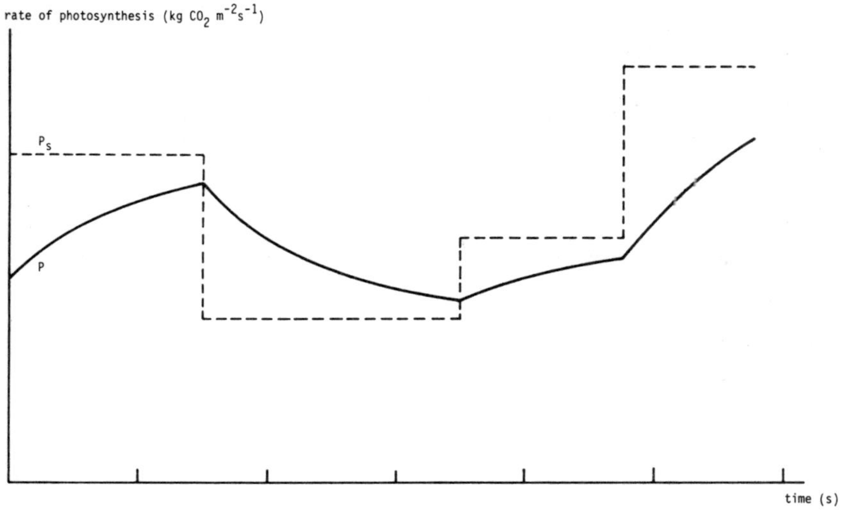

Exercise 9

a.

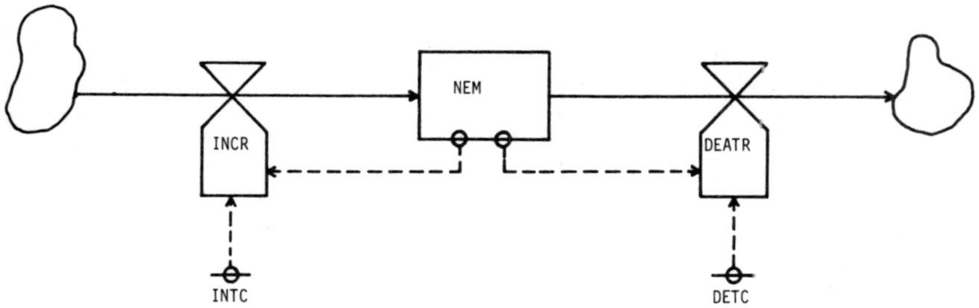

b. $NEM_t = NEM_{t-1} + DELT \times (INCR_{t-1} - DEATR_{t-1})$
$ININEM = 1000$
$INCR_t = 1/1 \times NEM_t$
$DEATR_t = \frac{1}{4} \times NEM_t$.

c. After 1 year (INCTC = 1) the number of nematodes has become e-times as large. After the next 4 years (DEATTC = 4), the number of nematodes has become e-times as small. Increase and decrease are cancelled within a 5-year crop rotation. The changes as function of time are sketched in the next figure.

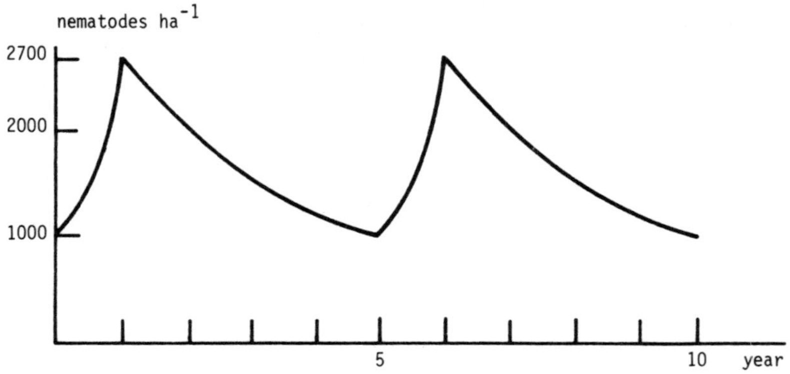

Exercise 10

a.

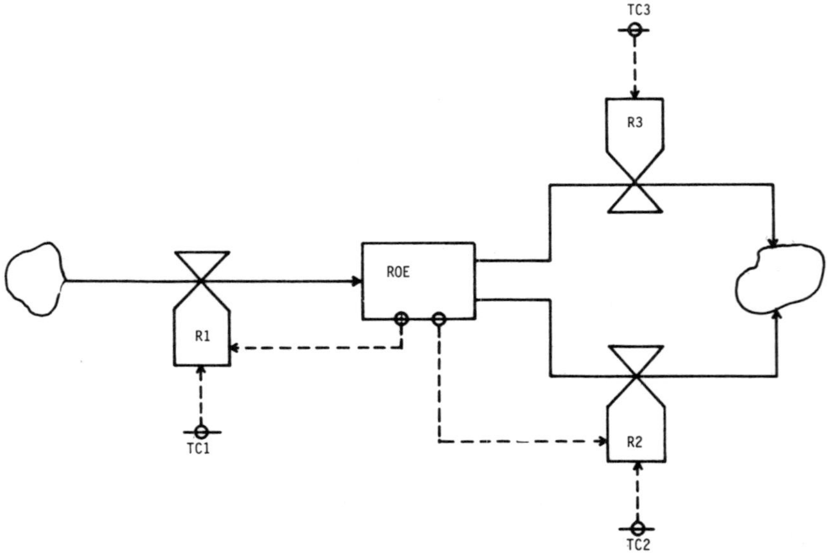

b. $ROE_t = ROE_{t-1} + DELT \times (R1_{t-1} - R2_{t-1} - R3_{t-1})$

$R1_t = \frac{1}{8} \times ROE_t$

$R2_t = \frac{1}{10} \times ROE_t$

The number of annual permits R3 is the unknown factor.

c. In a steady state with $ROE_t = 400$, the following equation applies:

$R1_t = R2_t + R3_t$.

Substitution gives:

$\frac{1}{8} \times ROE_t = \frac{1}{10} \times ROE_t + R3_t$ or $R3_t$ must be 10 roes a year.

d. It is an unrealistic model as the steady state can never be maintained in this problem. The system shows a positive feedback loop, by which every random discrepancy with the 400 roes effectuates a positive or negative growth (see also Exercise 5.3).

Exercise 11

a.

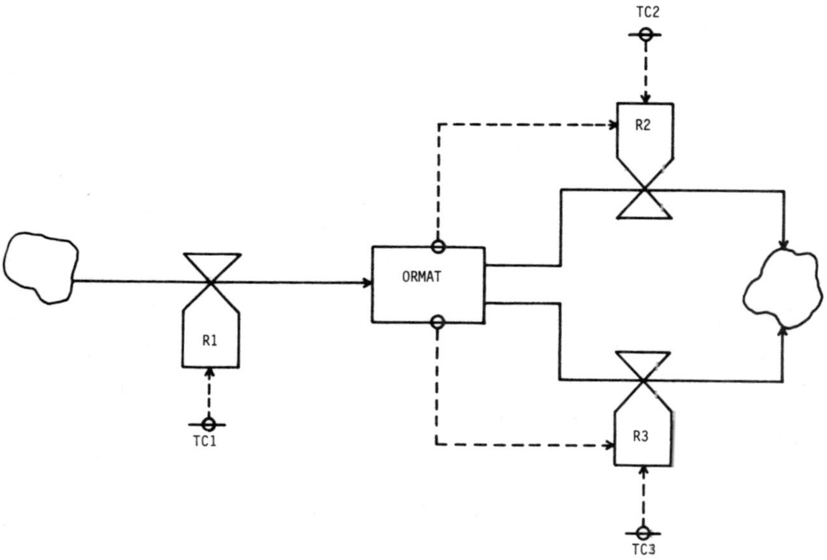

b. $ORMAT_t = ORMAT_{t-1} + DELT \times (R1_{t-1} - R2_{t-1} - R3_{t-1})$
$R1_t = 4000$ kg per ha
$R2_t = \frac{1}{50} \times ORMAT_t$
$R3_t = \frac{1}{200} \times ORMAT_t$
c. In the steady state:
$R1_t = R2_t + R3_t$.
Substitution gives:
$4000 = \frac{1}{50} \times ORMAT_t + \frac{1}{200} \times ORMAT_t$ or
$ORMAT_t = 160\,000$ kg per ha as the equilibrium value.

Exercise 12

a.

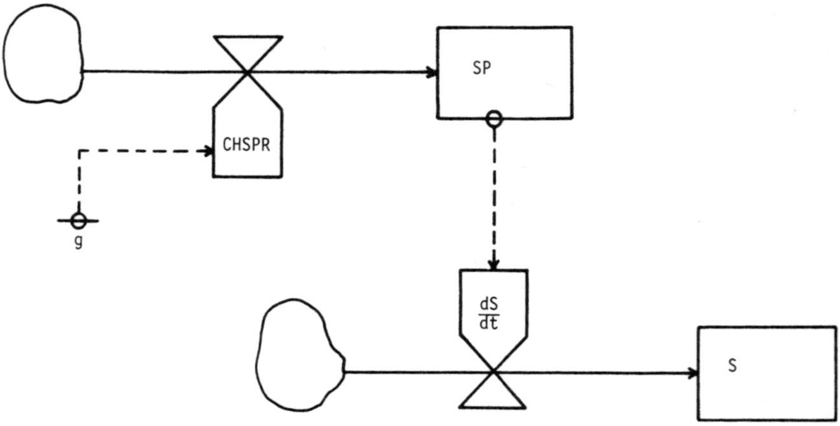

b. Call the rate of change of speed, the speed itself and the distance covered CHSPR, SP and S, respectively. Now according to the equations of free fall:

$SP_t = SP_{t-1} + DELT \times CHSPR_{t-1}$

$CHSPR_t = g$

$S_t = S_{t-1} + DELT \times SP_{t-1}$.

c. This system does not contain any feedback loop.

Exercise 13

a.

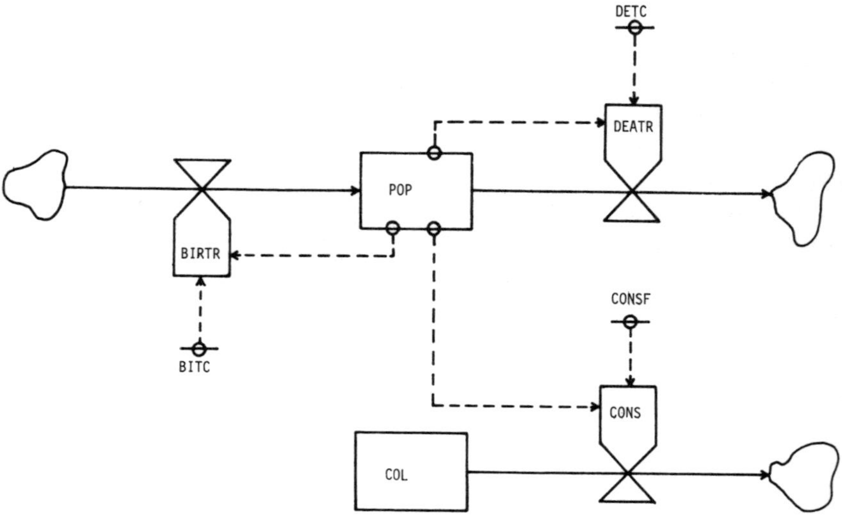

b. The increase by birth is 2% in the first year; as the relative birth rate remains constant, the time coefficient BTC equals 50 years. The time coefficient DTC of the death rate DEATR equals 65 years and the consumption factor CONSF to 1 ton person^{-1} year^{-1}.

$POP_t = POP_{t-1} + DELT \times (BIRTR_{t-1} - DEATR_{t-1})$
INIPOP = 5 000 000 inhabitants
$BIRTR_t = \frac{1}{50} \times POP_t$
$DEATR_t = \frac{1}{65} \times POP_t$
$COL_t = COL_{t-1} - DELT \times CONSF \times POP_{t-1}$
INICOL = 5000 million tons.

Exercise 14

a.

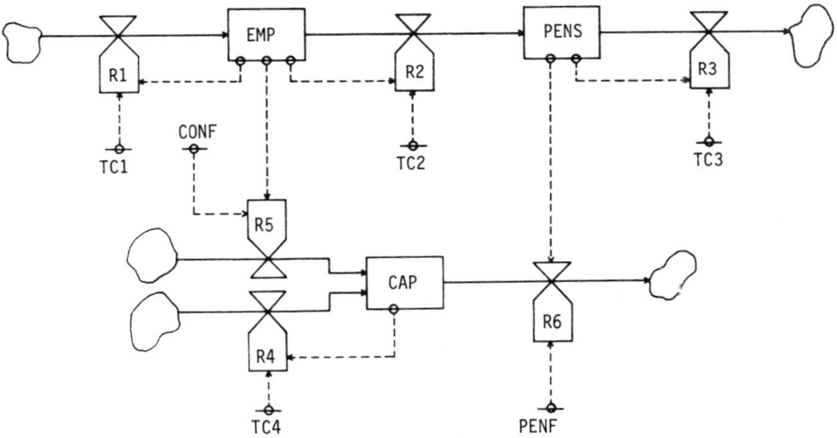

b. It appears from the data, that TC1 = 10 years, TC2 = 12.5 years, TC3 = 5 years, TC4 = 20 years, CONF = 500 guilders employee^{-1} year^{-1} and PENF = 10 000 guilders pensioner^{-1} year^{-1}.

$EMP_t = EMP_{t-1} + DELT \times (R1_{t-1} - R2_{t-1})$
INEMP = 300 persons
$PENS_t = PENS_{t-1} + DELT \times (R2_{t-1} - R3_{t-1})$
INPENS = 0 persons
$CAP_t = CAP_{t-1} + DELT \times (R4_{t-1} + R5_{t-1} - R6_{t-1})$
INCAP = 1 000 000 guilders
$R1_t = \frac{1}{10} \times EMP_t$
$R2_t = 1/12.5 \times EMP_t$
$R3_t = \frac{1}{5} \times PENS_t$
$R4_t = \frac{1}{20} \times CAP_t$
$R5_t = 500 \times EMP_t$
$R6_t = 10\,000 \times PENS_t$.

Exercise 15

a.

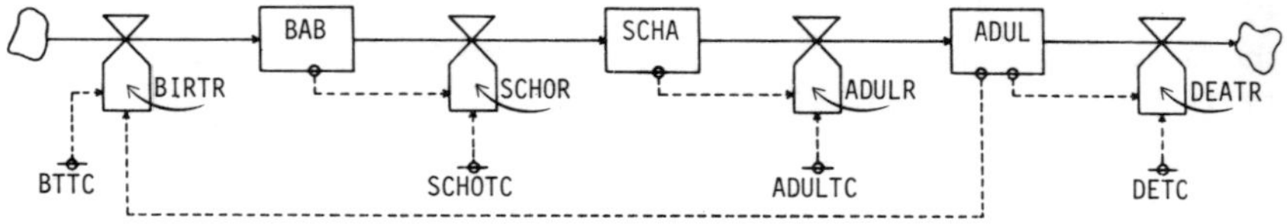

b. $BAB_t = BAB_{t-1} + DELT \times (BIRTR_{t-1} - SCHOR_{t-1})$
$SCHO_t = SCHO_{t-1} + DELT \times (SCHOR_{t-1} - ADULR_{t-1})$
$ADUL_t = ADUL_{t-1} + DELT \times (ADULR_{t-1} - DEATR_{t-1})$
$INBAB = 300$ $INSCHO = 3000$ $INADUL = 100\,000$
$BITC = 20$ $SCHOTC = 6$ $ADUTC = 10$ $DETC = 50$
$BIRTR_t = \frac{1}{20} \times ADUL_t$
$SCHOR_t = \frac{1}{6} \times BAB_t$
$ADULR_t = \frac{1}{10} \times SCHO_t$
$DEATR_t = \frac{1}{50} \times ADUL_t$.
c. Three negative feedback loops and one positive one.

Exercise 16

a.

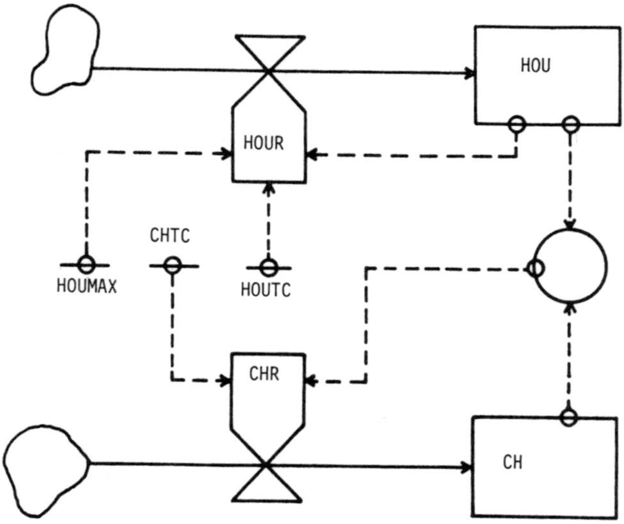

b. $HOU_t = HOU_{t-1} + DELT \times HOUR_{t-1}$
$HOUR_t = 1/HOUTC \times (HOUMAX - HOU_t)$
$CH_t = CH_{t-1} + DELT \times CHR_{t-1}$
$CHR_t = 1/CHTC \times (HOU_t - CH_t)$.

c.

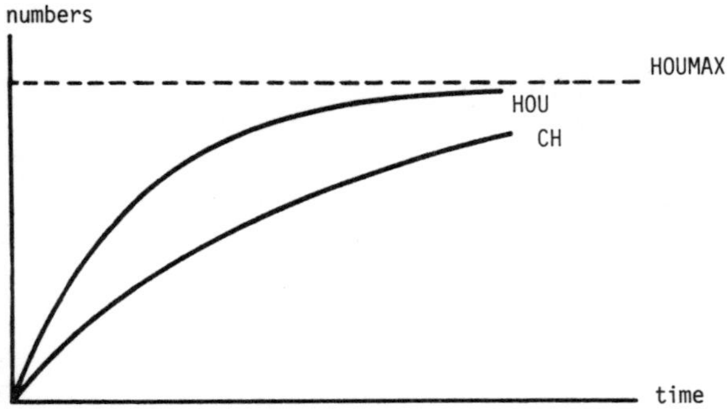

Exercise 17

The data show that the relative growth rate and therefore the time coefficient are not constant and influenced during the process by the amounts of nutrient in the solution; the nutrient uptake decreases the relative growth rate (or increases the time coefficient). Further, A is expressed in terms of the nutrient; in the relational diagram a direct flow from N to A can be drawn.

a.

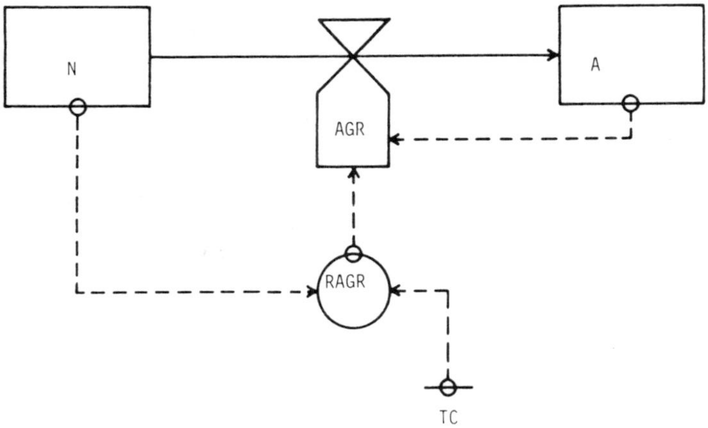

b. $N_t = N_{t-1} - DELT \times AGR_{t-1}$

$A_t = A_{t-1} + DELT \times AGR_{t-1}$

$AGR_t = 1/TC \times A_t \times N_t$.

c. month gram meter^{-3}

d. A negative feedback loop connecting N through RAGR and AGR with N, a positive one from A through AGR to A.

e. Call the initial amount of the nutrient ININ; then the amount of the nutrient N_t can be expressed as amount of A_t in grams N per m^3

according to $N_t = ININ - A_t$. Substituting this value in the rate equation gives:

$AGR_t = 1/TC \times (ININ - A_t) \times A_t$ or

$AGR_t = 1/TC \times ININ \times (1 - A_t/ININ) \times A_t$.

This is the equation of the logistic growth curve. The curves of A and AGR as functions of time are given in the figures of the answer to Question 20. The curve of N is the curve of A, reversed with respect to a line parallel to the time axis.

Exercise 18

a.

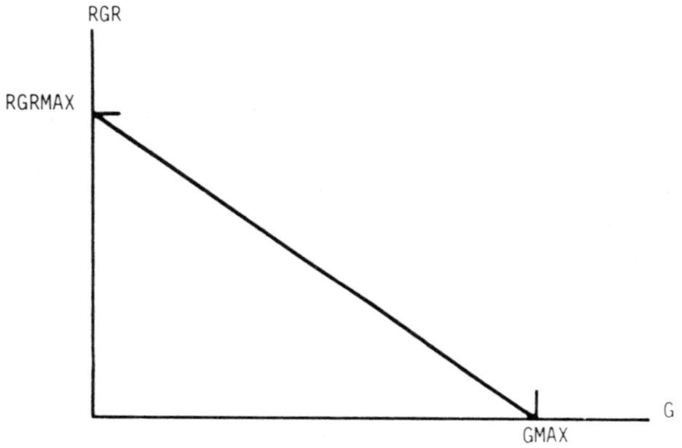

b. RGR = RGRMAX × (1 − G/GMAX) according to $y = m \times x + q$ in which m = −RGRMAX/GMAX.

c. It follows from the equation given in b that $dG/dt = G \times RGRMAX \times (1 - G/GMAX)$ or expressed with the time coefficient $dG/dt = G \times 1/(1/RGRMAX) \times (1 - G/GMAX)$.

This is the equation of the logistic growth curve of Question 20 in the text.

d. See the answer to Question 20.

e. See the answer to Question 20.

Exercise 19

a. It is clear from the description of the model assumed, that the time coefficient STITC (period in which one owner gives one stimulus) is constant in this case. The number of owners OWN affects the increase of owners directly and also indirectly through the effect of the stimuli.

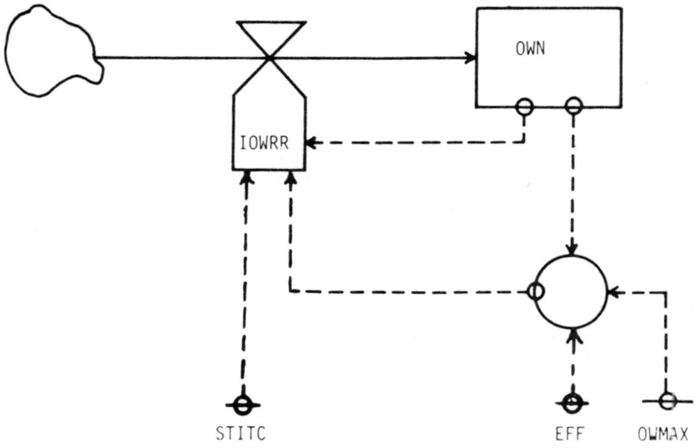

b. The rate of increase is the reciprocal of the time coefficient multiplied by the number of stimuli per owner, by the number of owners and by the effectivity EFF of the stimuli. However, this effectivity is influenced by the dimensionless factor $(1-OWN_t/OWMAX)$. The dimension of the right side of the rate equation is equal to: owner time^{-1}. The rate equation itself is: $IOWRR_t = 1/STITC \times OWN_t \times EFF \times (1-OWN_t/OWMAX)$. This equation is comparable with that of the logistic growth of yeast. Therefore, the model gives an explanation of the S-shaped curve of the number of owners as a function of time.

Exercise 20

a.

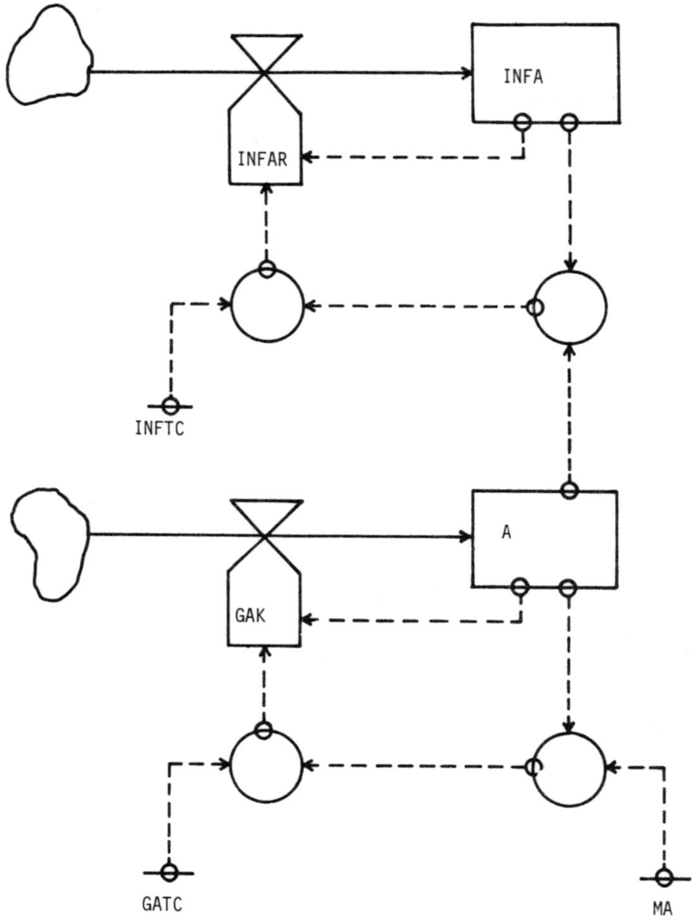

b. $INFA_t = INFA_{t-1} + DELT \times INFAR_{t-1}$

$A_t = A_{t-1} + DELT \times GAR_{t-1}$

$INFAR_t = INFA_t \times 1/INFTC \times ((A_t - INFA_t)/A_t)$

$GAR_t = A_t \times 1/GATC \times (1 - A_t/MA)$.

This system can be described by a differential equation: $dy/dt = a \times y \times (k(t) - y)$ in which $k(t) = k/(1 + e^{-m \times t})^{1/m}$; the integral of this differential equation is very complex (in these equations other symbols are used).

Exercise 21

a. It appears from the description of the problem that the system contains a second state variable with initial value M_0 besides the variable W. Both differential equations become: $dW/dt = M \times W$ in which 1/M is the time coefficient, and $dM/dt = -S \times M$ with 1/S as time coefficient.

b.

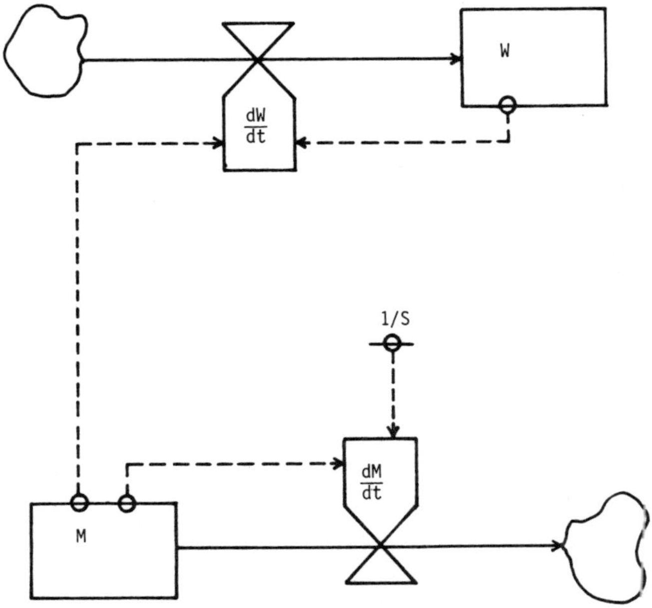

c. As time goes on, the factor M of the first equation becomes smaller according to the process described by the second equation. Therefore, the growth curve is nearly logistic.

Exercise 22

a. W(t)/V, is the pollution load of the lake, $-c/t_0$ the reduction of the organic matter by discharge and $-k \times c$ the reduction by decomposition.

b. The mean residence time is t_0.

Exercise 23

a.

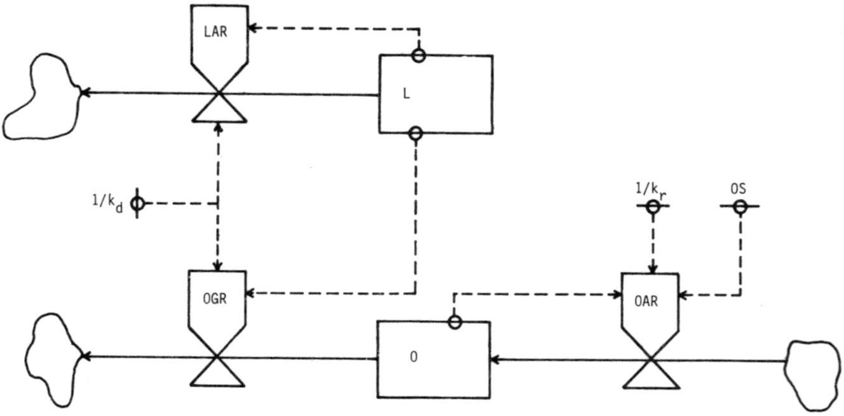

b. $LAR_t = k_d \times L_t$

$OGR_t = k_d \times L_t$

$OAR_t = k_r \times (OS - O_t)$

$O_t = O_{t-1} + DELT \times (OAR_{t-1} - OGR_{t-1})$

$L_t = L_{t-1} - DELT \times LAR_{t-1}$

in which LAR, OGR and OAR are the decomposition rate of the organic waste, the rate of oxygen depletion used for this decomposition and the reaeration rate, respectively. OS is the saturation value of oxygen in water at specific conditions.

c. By the decomposition of the organic waste, a corresponding loss of oxygen in water takes place; according to the decomposition equation, this loss is exponential. The actual reduction of oxygen in water is less as reaeration takes place, though delayed; the oxygen content of the water continues decreasing until reaeration has been become equal to the loss of oxygen by decomposition. After this critical point, the oxygen content of the water will increase and reach ultimately the saturation value OS.

Exercise 24

a.

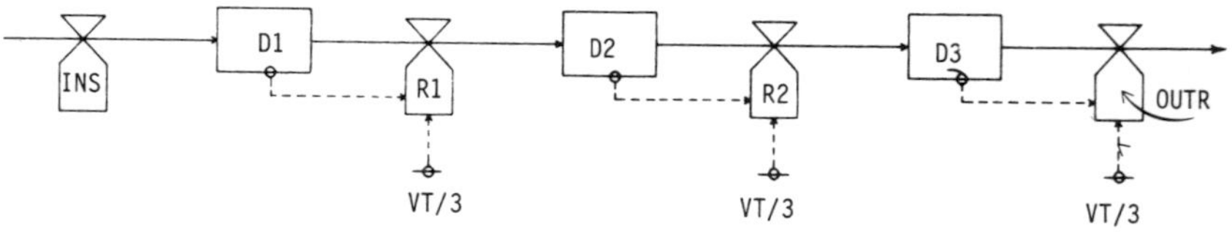

b. In the steady state with a total delay time VT, with an exponential delay of the third order and with an input INS, the amounts entering each delay element D are equal to the amounts leaving. It follows that the content of an element just before the change is INS×(VT/3). The equations are:

$D1_t = D1_{t-1} + DELT \times (INS_{t-1} - R1_{t-1})$

$R1_t = 1/(VT/3) \times D1_t$

$D2_t = D2_{t-1} + DELT \times (R1_{t-1} - R2_{t-1})$

$R2_t = 1/(VT/3) \times D2_t$

$D3_t = D3_{t-1} + DELT \times (R2_{t-1} - OUTR_{t-1})$

$OUTR_t = 1/(VT/3) \times D3_t$.

c.

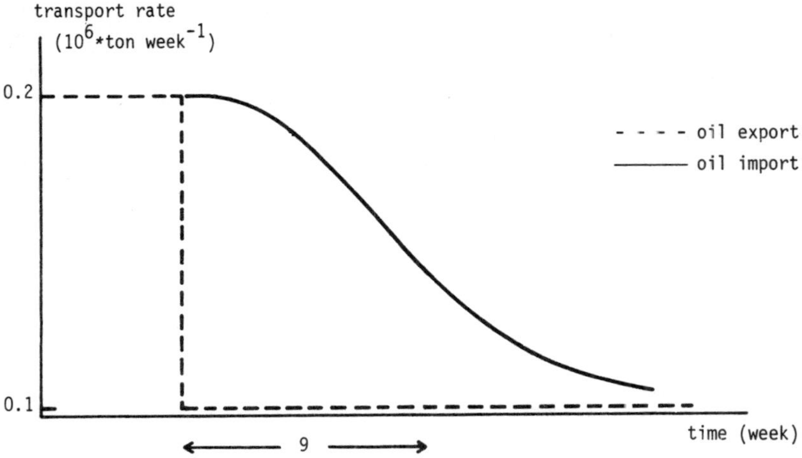

Exercise 25

a.

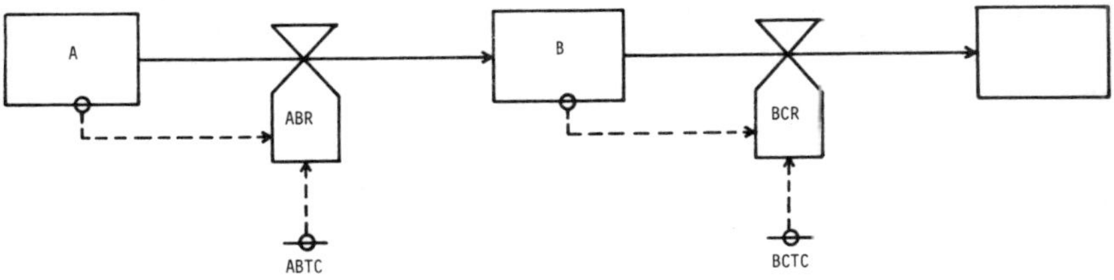

b. $A_t = A_{t-1} - DELT \times ABR_{t-1}$

$B_t = B_{t-1} + DELT \times (ABR_{t-1} - BCR_{t-1})$

$C_t = C_{t-1} + DELT \times BCR_{t-1}$

INIA = 100 INIB = 0 INIC = 0

$ABR_t = 1/ABTC \times A_t$

$BCR_t = 1/BCTC \times B_t$.

c. The curve of A is always decreasing exponentially. In the first calculation interval, no C is produced as there is no B yet; during this interval, the curve of B is the reverse of the curve of A and convex upwards. In the next interval, B always increases less than A decreases because of the reduction of B by the simultaneous conversion into C. The amounts of B reach a maximum value when the increase in B by conversion of A equals the decrease in B by the production of C. In the first intervals, hardly any C is produced, the shape of the C curve is concave upwards with an ascending slope. This slope reaches a maximum or point of inflexion at the moment that B has its maximum value. Therefore, the curve of C is S-shaped and attains its maximum, when there is no A left and all B is converted into C.
The ratio ABTC/BCTC determines, whether the maximum of B lies above or under the A curve. For $1/ABTC \times A_t = 1/BCTC \times B_t$ or $BCTC/ABTC = B_t/A_t$ applies to the maximum point of B. The coefficient ABTC is larger than BCTC at a relatively slow production of B and then the maximum of B lies under the A curve, and conversely. The behaviour pattern of the system is given in the following figure.

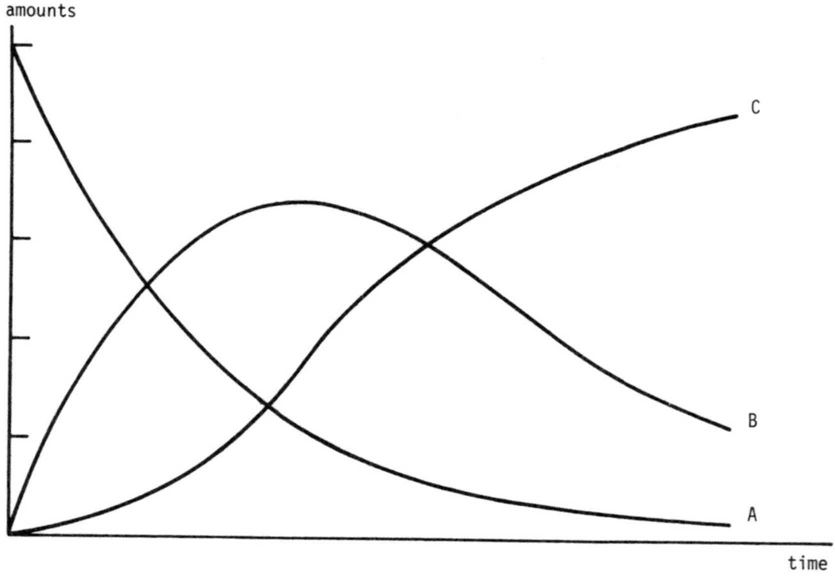

This exercise gives an insight into the working of successive integrations on the response patterns of delays. It makes clear, that an increasing number of integrations or the order of delay enlarges the time before the last integration reacts to the change in the input: with a delay of infinite order, the output reacts just after the total delay time, which yields the pipeline effect.

Exercise 26

a.

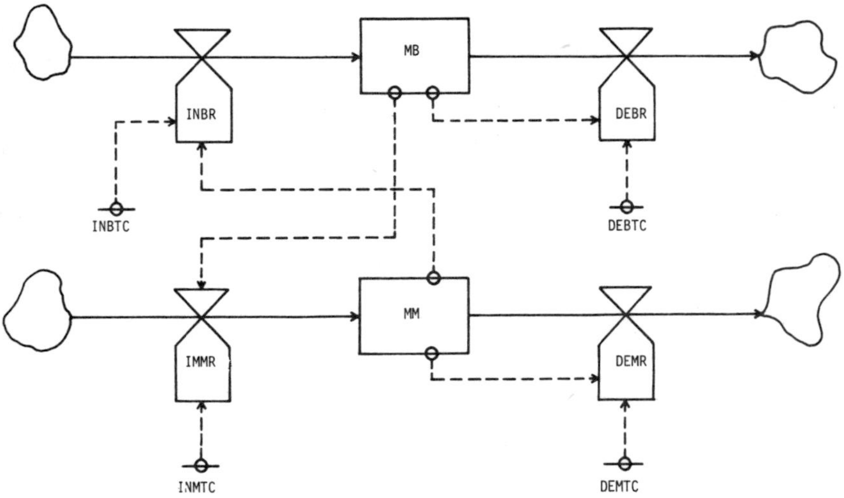

b. $MB_t = MB_{t-1} + DELT \times (INBR_{t-1} - DEBR_{t-1})$
$MM_t = MM_{t-1} + DELT \times (INMR_{t-1} - DEMR_{t-1})$
$INBR_t = 1/INBTC \times MM_t$
$INMR_t = 1/INMTC \times MB_t$
$DEBR_t = 1/DEBTC \times MB_t$
$DEMR_t = 1/DEMTC \times MM_t$.

c. Two negative feedback loops of the depreciation rates of B and M; a large positive loop running from B through INMR, MM and INBR to MB.

d. In a steady state, the rates of inflow must be equal to the rates of outflow. Therefore, $INBR_t = DEBR_t$ and $INMR_t = DEMR_t$. Substitution gives:
$1/INBTC \times MM_t = 1/DEBTC \times MB_t$ and
$1/INMTC \times MB_t = 1/DEMTC \times MM_t$.
Further working-out shows, that a steady state will be attained, if $1/INMTC \times 1/INBTC = 1/DEBTC \times 1/DEMTC$.

e. The necessary conditions for an explosive growth can be derived from the results of d: the products of the time coefficients of investments must be smaller than those of depreciation. In the next figure, the investments of both firms as function of time are given for the three cases:
$INMTC \times INBTC \gtreqless DEBTC \times DEMTC$.

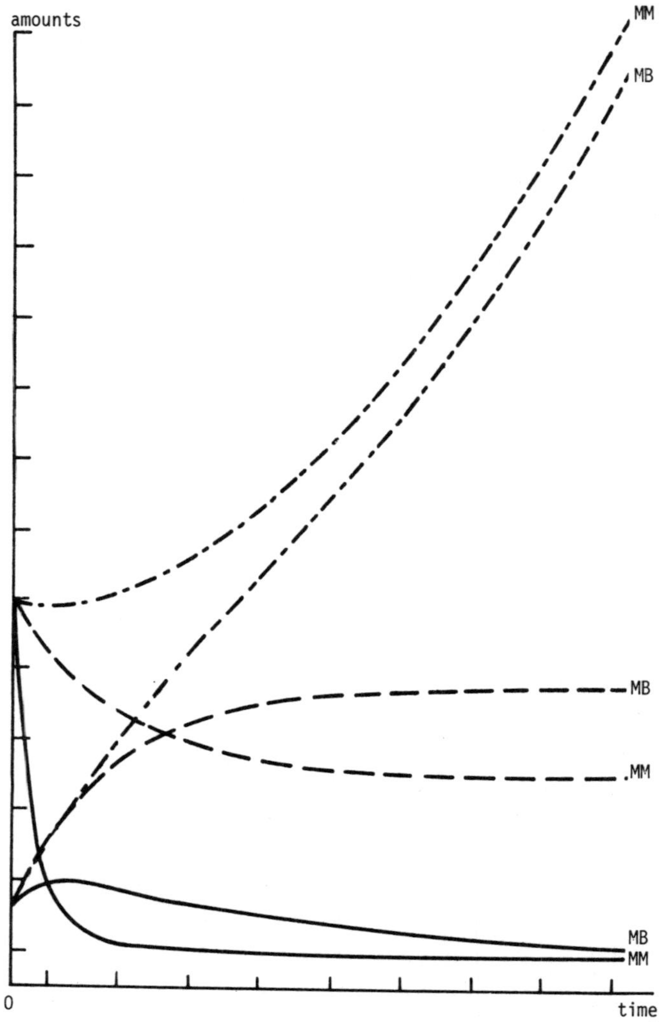
amounts
MM
MB
MB
MM
MB
MM
0
time

Exercise 27

a.

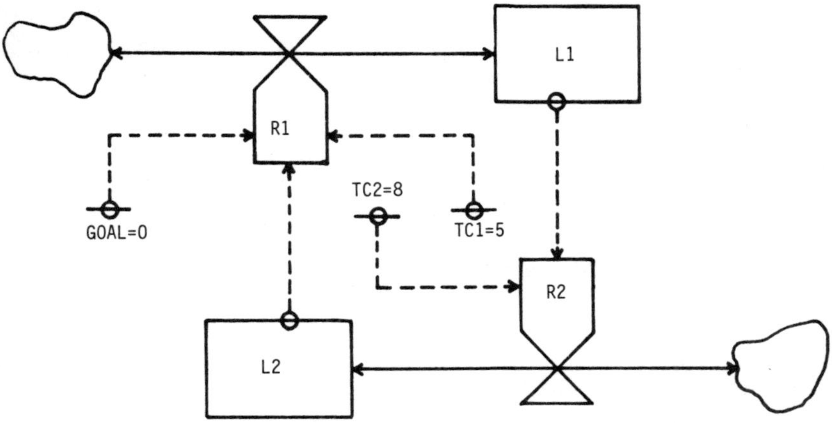

b. A negative feedback loop from R1, through L1, R2 and L2 to R1.

c.

moment	change in L1	L1	change in L2	L2	R1	R2
0		10.00		0.00	0.00	1.25
2	0.00	10.00	2.50	2.50	−0.50	1.25
4	−1.00	9.00	2.50	5.00	−1.00	1.13
6	−2.00	7.00	2.25	7.25	−1.45	0.88
8	−2.90	4.10	1.75	9.00	−1.80	0.51
10	−3.60	0.50	1.03	10.03	−2.01	0.06
12	−4.01	−3.51	0.13	10.15	−2.03	−0.44
14	−4.06	−7.57	−0.88	9.27	−1.85	−0.95
16	−3.71	−11.28	−1.89	7.38	−1.48	−1.41
18	−2.95	−14.23	−2.92	4.56	−0.91	−1.78
20	−1.82	−16.06	−3.56	1.00	−0.21	−2.01

The ultimate result is the characteristic behaviour pattern of two integrations connected with each other in a negative feedback loop, represented by a differential equation of the second order. This pattern consists of oscillations as given in Fig. 13 of the text. These oscillations are not damped and the equilibrium state will never be attained permanently. Notice the shifts or delays between L1 and L2 etc. However, a system with two integrations, connected with each other in a positive feedback loop, exhibits an explosive growth. Some examples are: the system describing the motion of a relatively long pendulum with L1 and L2 as speed and position of pendulum end, respectively, the placing of orders L1 dependent on the stock L2 compared with a standard stock (=goal or equilibrium), the Lotka-Volterra equations about the interaction between host H and parasite P: $\mathrm{d}H/\mathrm{d}t = (\mathrm{a} - \mathrm{b} \times P) \times H$ and $\mathrm{d}P/\mathrm{d}t = (-\mathrm{c} + \mathrm{e} \times H) \times P$ (compare these feedback loops with those of the equations of Exercise 29).

Exercise 28

a. See Fig. 12 in the text, in which L2 and L1 are P and v respectively; the goal or equilibrium equals to zero.

b. Call $\mathrm{d}v/\mathrm{d}t$, v and the rate of change of position ACC, V and PCR respectively. The equations concerned are now:

$\mathrm{ACC}_t = -\mathrm{g}/1 \times \mathrm{P}_t$

$\mathrm{V}_t = \mathrm{V}_{t-1} + \mathrm{DELT} \times \mathrm{ACC}_{t-1}$

$\mathrm{PCR}_t = \mathrm{V}_t$

$\mathrm{P}_t = \mathrm{P}_{t-1} + \mathrm{DELT} \times \mathrm{PCR}_{t-1}$.

c. It is a system with two integrations, connected with each other in a negative feedback loop. The behaviour pattern is described by sine and cosine functions as given in Fig. 13 of the text.

Exercise 29

a.

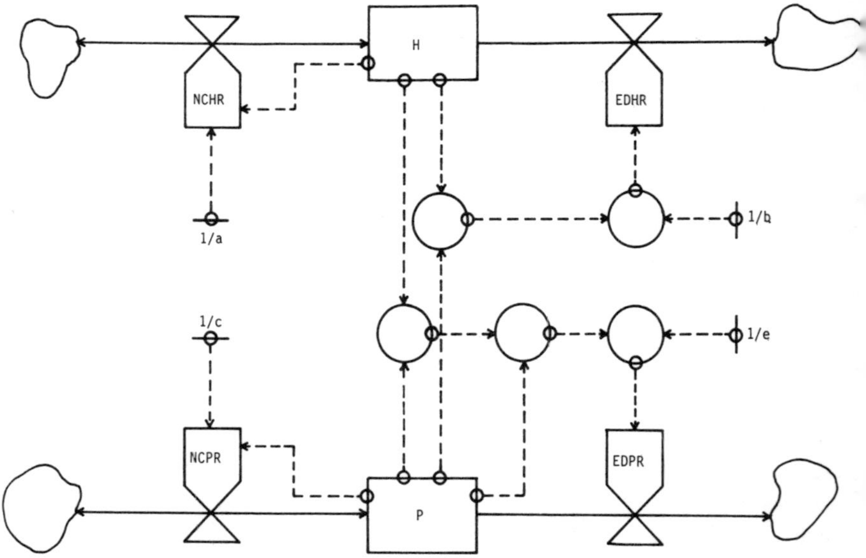

b. $\mathrm{H}_t = \mathrm{H}_{t-1} + \mathrm{DELT} \times (\mathrm{NCHR}_{t-1} - \mathrm{EDHR}_{t-1})$

$\mathrm{P}_t = \mathrm{P}_{t-1} + \mathrm{DELT} \times (\mathrm{NCPR}_{t-1} - \mathrm{EDPR}_{t-1})$

$\mathrm{NCHR}_t = 1/(1/\mathrm{a}) \times \mathrm{H}_t$

$\mathrm{EDHR}_t = 1/(1/\mathrm{b}) \times \mathrm{H}_t \times \mathrm{P}_t$

$NCPR_t = 1/(1/c) \times P_t$
$EDPR_t = 1/(1/e) \times P_t \times P_t/H_t$.
c. The steady state will be attained, when $a \times H_t = b \times H_t \times P_t$ and $c \times P_t = e \times P_t \times P_t/H_t$. Then $P = a/b$ and $H = a \times e/(b \times c)$.
d. Two positive feedback loops determine the natural net-increase of P and H. There is also a large negative feedback loop ranging from P through EDHR, H, EDPR to P. Two smaller negative feedback loops are present within this large loop, ranging from P through EDPR to P and from H, through EDHR to H, respectively. The behaviour pattern of such systems is represented by oscillations which may be damped to an equilibrium or steady state. Compare this behaviour with that of a system as described by the Lotka–Volterra equations, discussed in the solution to Exercise 27.

Exercise 30

a. Let the mean rate of commissioning be BUIR, the mean depreciation rate DER, the relationship between mean annual turnover per restaurant and the number of road restaurants FUNC, the percentage of the total turnover set apart for maintenance of and building new restaurants PERC and the building costs per restaurant ORCF. The relational diagram of this system with a delay of the first order is given in the next figure.

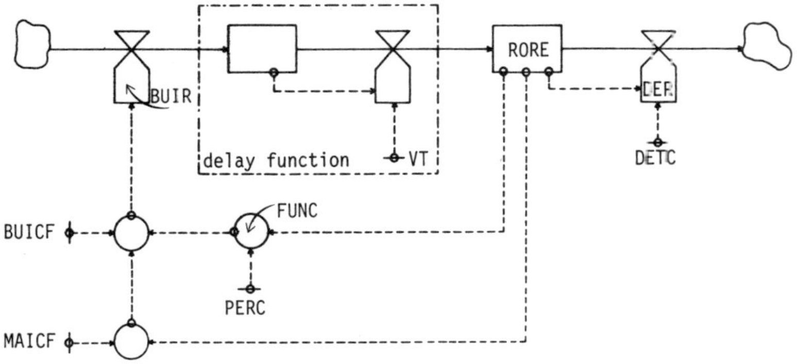

b. $RORE_t = RORE_{t-1} + DELT \times (BUIR_{t-1} - DER_{t-1})$
$INIROR = 100$
$BUIR_t = (RORE_t \times TUO_t \times PERC - MAICF \times RORE_t)/BUICF$
$DER_t = 1/DETC \times RORE_t$
DETC 20 VT = 3 MAICF = 100 000 PERC = 0.025
BUICF = 2 500 000; FUNC is given.
After substitution, the rate equations are:
$BUIR_t = (RORE_t \times TUO_t \times 0.025 - 100\,000 \times RORE_t)/2\,500\,000$
$DER_t = \frac{1}{20} \times RORE_t$.

c. The dimensions of the right side of the first rate equation are: restaurant × (guilder/(year × restaurant)-guilder/(year × restaurant))/(guilder/restaurant). Working out gives: restaurant/year which is also the dimension of the left side.

d. A reconstruction of the rate equation BUIR gives: $BUIR_t = (RORE_t \times (TUO_t/40 - 100\,000))/2\,500\,000$. Thus the system is governed by a positive feedback loop so long as TUO/40 is larger than 100 000. The positive feedback will be changed into a negative one, when the mean turnover is smaller than 4 000 000; this change takes place when the number of restaurants exceeds 125.

References

Boulding, K. E., 1956. Toward a general theory of growth. General Systems, 1: 66–75.

Ferrari, Th. J., 1965. Prüfung mit und ohne Eingriff. In: Aktuelle Probleme des landwirtschaftliche Versuchswesens. Landwirtschaftliche-Chemischen Bundesversuchsanstalt, Linz, Veröffentlichungen 6: 25–32.

Forrester, J. W., 1961. Industrial dynamics. Massachusetts Institute of Technology Press, Cambridge, Mass. U.S.A.

Forrester, J. W., 1968. Principles of systems. Wright-Allen Press, Cambridge, Mass. U.S.A.

Goudriaan, J., 1973. Dispersion in simulation models of population growth and salt movement in the soil. Neth. J. Agric. Sci 21: 269–281.

Kuo, S. S., 1965. Numerical methods and computers. Addison-Wesley, Reading, U.K.

Meadow, D. L. et al., 1972. The limits to growth. Universe Books, New York, U.S.A.

Shannon, R. E., 1975. Systems simulation: the art and science. Prentice-Hall, Englewood Cliffs.

Wit, C. T. de & Goudriaan, J., 1978. Simulation of ecological processes. Pudoc, Wageningen, the Netherlands.

Index

2913